Silent Winter

SILENT WINTER

OUR CHEMICAL WORLD
AND
CHRONIC ILLNESS

Joanna Malaczynski-Moore

Algora Publishing
New York

Library of Congress Cataloging-in-Publication Data

Names: Malaczynski-Moore, Joanna, 1977– author.
Title: Silent winter: our chemical world and chronic illness / by Joanna
    Malaczynski-Moore.
Description: New York: Algora Publishing, [2021] | Includes
    bibliographical references and index. | Summary: "*Silent Winter* is
    about the spread of toxic chemicals in our lives and their role in the
    growing prevalence cancer, chronic fatigue, diabetes, asthma digestive
    issues, depression, dementia, and other diseases. Scientific evidence
    about chronic illness and toxic chemicals is withheld from the public
    through stunningly elaborate efforts so that business can continue as
    usual."— Provided by publisher.
Identifiers: LCCN 2020057078 (print) | LCCN 2020057079 (ebook) | ISBN
    9781628944471 (hardcover) | ISBN 9781628944464 (trade paperback) | ISBN
    9781628944488 (pdf)
Subjects: LCSH: Environmental toxicology. | Environmentally induced
    diseases. | Environmental health.
Classification: LCC RA1226 .M35 2021 (print) | LCC RA1226 (ebook) | DDC
    613/.1—dc23
LC record available at https://lccn.loc.gov/2020057078
LC ebook record available at https://lccn.loc.gov/2020057079

Cover photo by Hugo Jehanne on Unsplash.

Printed in the United States

TABLE OF CONTENTS

Introduction

Sometimes you make your own plans in life and sometimes life has its own plans for you. In 2014, I started swimming in the Willamette and Columbia Rivers outside of town because I just could not get to the ocean enough. These waterways are fairly urbanized and heavily polluted, but I couldn't resist—even though my professional work focused on eliminating toxic chemicals from our environment.

The rivers were contaminated with highly persistent, bio-accumulative, and toxic chemicals. Upstream were pesticide-laden agricultural fields. Motor boats regularly passed by with their polluting two-stroke engines. And tanker ships hovered in the distance. But somehow I did not think any of this would impact me. I wanted to have the freedom to be in open waters so badly that I was willing to turn a blind eye to the pollution around me.

Within a short while, I started to get sick. I came down with the worst flu-like symptoms ever. I was exhausted for months, my head and sinuses were killing me, and my stomach was heavily bloated. There were days I mostly slept, assuming I would get over it. Other days I worked and even traveled. Doctors tested me for various infections, including viruses and parasites. In the end, I received no help from the conventional medical establishment; they told me I was too healthy to be ill. Somehow I managed through this time with the help of acupuncture and Chinese medicine.

This went on until I ended up at the hospital for intestinal surgery; I was diagnosed with an intestinal infection. Maybe I picked up something in the polluted waters. Maybe it was something else.

But surgery was not the end of my problems. The fatigue and bloating continued. The sinus problems became even worse. I could barely breathe through my nose. This caused me to make the rounds again, seeking medical help. This again went nowhere.

I continued to believe that I was the victim of some unresolved infection. But I was opening up to the idea that other forces were at play. My acupuncturists looked at my problems differently—they thought the issues had to do with the overall health of my body and immune system. Moreover, the intestinal surgeon pointed out to me that the human body—including the intestines—are full of bacteria, viruses, and other organisms, and yet they do not normally attack us.

I started to reflect upon my life for other causes of these ailments. That is when it occurred to me that I was pushing my luck swimming in polluted waters, and that maybe I had pushed my body over the edge. Over the years I had become more chemically sensitive. A morning trip to any industrial facility—with its inevitable poor air quality and largely ineffective ventilation system—would usually set me off. I would become very tired by the afternoon, lie down, and not be able to get back up until the next morning. It was clear that these sudden onsets of fatigue were chemically related.

The connection between chronic illness and chemical exposure was made even more obvious in the spring of 2019. On a vacation to Mexico, I was exposed to a high dose of toxic chemicals—this time in the form of fragranced cleaning products, which saturated our hotel sheets, mattress, towels, and even toilet paper. The smell persisted, soaking into our clothes and skin, despite our efforts to air out the bedding. The hotel would not help, telling us that their laundry was done off site.

By day three in this environment, I began to feel a deep pain in my uterus that I can only describe as pelvic inflammation. This alarmed us both and we decided it was time to leave. We went so far as to check out of the hotel and go to the airport, only to find out we would not be able to fly home that day. We returned to town distressed, but ultimately found another hotel that did not use much fragrance.

We thought we had dodged a bullet. The pain in my uterus subsided and we seemed to be functioning well enough when we got home. But the tee-shirts I slept in at the original hotel ended up permanently smelling like fragrance. I washed them over a dozen times and left them to air out for days. Unfortunately the scent was bound to the fabric by problematic chemicals that are resistant to sunlight, detergent, and even enzyme cleaner. No matter how many times I washed them or hung them outside, their smell persisted.

The toxic chemicals from this fragrance persisted in our bodies as well. My husband was the first to notice that something was off. He started experiencing cyclical and very bothersome mood swings. Then I noticed rising irritability as well. Weeks later I began struggling with acute diabetic-like symptoms. I got sicker than my husband because I had significantly exposed myself to persistent toxic chemicals in recent years and they were still present in my body.

The resulting cocktail of chemicals was interfering with my ability to effectively utilize the glucose from the food I had eaten. I was not absorbing it from my bloodstream into my cells. It left me catatonic. For the next several months I avoided all sugars, fruits, and even grains. I took Chinese herbs for months and went to acupuncture twice a week. Nevertheless, I was fatigued, constantly hungry, and sometimes barely lucid.

My sense of smell and reaction to airborne toxins was extremely heightened and I became very intolerant of chemicals. I was bothered by fragranced laundry detergent on other people's clothes. I could no longer use any of my fragranced shampoos or lotions, and even became intolerant to the smell of plastics. Any exposure to these would send me reeling into a state of fatigue and hibernation that could last for days.

In order to cope, I had to make my own personal care and cleaning products, substituting organic cocoa butter and coconut oil for body lotions. Baking soda and vinegar became cleaning staples. I bought calcium carbonate and zinc oxide to make items such as toothpaste and sunscreen. And a friend taught me to use a lemon wedge as an effective deodorant (you rub the juice as well as the peel onto your skin).

Because of my intolerance to plastics, I had to replace most of my synthetic clothes with natural fibers, such as hemp and cotton. This included my bras, underwear, bathing suits, shirts, pants, and anything else next to my skin. I also stopped eating food packaged in plastic and eliminated personal care products packaged in plastic.

My symptoms mostly subsided over time, but I had to go fairly extreme in order to heal. I was already eating a healthy diet, limiting sugar, and eating organic produce. But that was not enough once I got sick. I had to secure clean meat, eggs, and dairy from local farmers. I could no longer eat corn, wheat or soy, unless I was 100% sure it was organic. As a result of all this, it was impossible for me to go out to a restaurant, because I could not verify the purity of the food. One year before COVID-19 hit, I was already in quarantine as a result of my experience with our chemical world.

My imperative to write this book became quite urgent. I originally thought I was going to write a book about chronic fatigue and chemical sensitivity, the symptoms that haunted me the most. The more I researched for this book, however, the more I learned that the vast majority of chronic illnesses—such as cancer, diabetes, intestinal disorders, skin disease, allergies, autoimmune disease, and mental illness—are on the rise as a result of toxic chemicals.

Around this time I started reading a well-known book: Rachel Carson's *Silent Spring*. On pages 187–198, Carson described the symptoms of toxic chemical poisoning.[1] It turned out that many of them matched my own. She addressed the liver damage caused by toxic chemicals, and how that damage affects the body's ability to deal with additional foreign substances. She also documented the fatigue, aching of limbs, and irritability that come with chemical exposure. And how many of the effects of toxic chemical poisoning can be delayed by weeks and even months after exposure, obscuring the chemical source of the illness.

Carson also warned of the health hazards of exposure to multiple toxic chemicals from our environment and the unpredictable effects these chemical mixtures can have on us. She drew attention to the poorly understood impacts that toxic chemicals have on the intricate metabolic processes in our bodies that keep us alive and functioning properly. Indeed, in most prophetic terms, Carson warned us in 1962 that toxic chemicals were an environmental time bomb.

But we did not listen. Nearly 60 years later, we are seeing the chronic health impacts of our widespread use of toxic chemicals. And 45% of the adult US population now has at least one chronic illness under the influence of our toxic chemical world.[2]

Most of us are told that our health issues are a result of our lifestyle or genes. We rarely hear that chronic illness is on the rise as a result of toxic chemicals in consumer products and throughout our environment. This information is withheld from us through some stunningly elaborate efforts. Highly sophisticated PR techniques are used to convince us that all is well and aggressive corporate marketing strategies are used to encourage us to buy toxic products we do not need. Industry knows toxic chemicals are a problem but does not want to change, so it is forcing us to change on an evolutionary level to deal with the onslaught of chemicals it has introduced into our daily lives. Inevitably our bodies cannot keep up and we get ill.

I wrote this book to make this clear and present danger apparent to you. You should know that chronic illness does not simply come out of the blue, nor is it predestined by your genes, nor should you blame it exclusively on

---

[1] Rachel Carson, *Silent Spring*, ch. 12 (The Riverside Press 1962).

[2] Wullianallur Raghupathi and Viju Raghupathi, "An Empirical Study of Chronic Diseases in the United States: A Visual Analytics Approach to Public Health," *International Journal of Environmental Research and Public Health* 15(3), 431 (March 1, 2018); individual health statistics for the US are also available from the US Centers for Disease Control, National Center for Health Statistics, available at https://www.cdc.gov/nchs/.

lifestyle or stress. I want you to understand that chronic illness is commonly caused by chronic exposure to toxic chemicals. That your exposure affects the health of future generations and that even the chemicals you were exposed to in the womb can affect your own unborn children. That industry has known about this for decades. And that your government, aware of this as well, has been totally ineffective in protecting your health.

Do not be disheartened by what I have to tell you. Use the knowledge I have compiled and shared with you to create a better world for yourself and those around you. If you are a victim of our chemical world, know that you are neither alone nor helpless. Millions of people suffer from the same health challenges as you do, and there is important strength in numbers. Also recognize that your personal experience is a powerful force and that you have the option of utilizing it to turn this ship around.

This book will help you along your journey. Chapter 1 paints a future toward which we are heading and frames our current health crisis. Chapters 2–3 introduce the disruption that man–made chemicals have caused in our world. To help explain how we got here, Chapters 4–7 discuss the economic and political forces at work that have enabled such large-scale pollution. Chapters 8–14 lay out specific examples of companies, chemicals, and products that need to be on your radar.

Chapters 15–22 explain how chemicals affect our bodies. Most chronic illnesses are tied to toxic chemicals, and this book will show how these chemicals cause various chronic illnesses. Some chronic illnesses are already acknowledged in many circles as "environmental illnesses," characterized by flu-like symptoms, chronic fatigue, and multiple-chemical sensitivity. Chapters 21–22 describe these illnesses and share some of the personal stories of individuals who suffer from them.

The final chapters of this book (Chapters 23–24) share my thoughts on the root causes of our environmental woes. You might expect that it's a lack of financing or technology that keeps us from creating a healthier and cleaner world. Technology and financing are never to blame, however; they are simply not a priority. Billions of dollars are spent by companies to deny the existence of the problem and to resist the implementation of sustainable technologies because the status quo suits them better. That same level of investment could instead be going to the individuals and companies already offering cleaner solutions.

There are two ways to get companies to change. One way is to light a fire in the hearts of corporate leadership by inspiring them to do something different. The other way is to light a fire under their butts by forcing them to change. I hope you will be a catalyst for one or both.

PART I: OUR CHEMICAL WORLD

CHAPTER 1: OUR CURRENT TRAJECTORY

> One day everything will be well; that is our hope. Everything's fine today; that is our illusion.
>
> — Voltaire

Imagine a world where you are completely dependent on pharmaceuticals. You take one set of drugs that protect your cells and DNA from damage caused by toxic chemicals. You take another set of drugs to keep your system detoxifying all of the synthetic chemicals used in your environment. And a third set of drugs to keep your immune system, hormones, and other systems functioning properly. If you get sick or your job exposes you to more toxic substances than your body can take, you get another set of drugs to help repair your body. If you were to stop any of these drugs, your body would degenerate and within weeks or months you would probably be dead.

Your drug regimen costs as much as your rent. You work in order to pay these bills and in order to live. You wonder what will happen if you get behind and fall into debt. Things will become even more expensive. Debt collectors will help you, but at a relatively steep interest rate. If you have good genes, you may be able to sell some of your DNA or other extracts from your body to cover your bills. If you are not so lucky, then you will have to work more. The most unpleasant jobs will expose you to more toxic substances. And they will result in you needing to buy more drugs. The costs of living this lifestyle keep escalating, further aggravating your anxiety.

Some people just give up and stop taking their drugs, with terrible consequences. At first they may develop strange flu- or allergy-like symptoms, such as sinus congestion, headaches, joint and muscle aches, fever, fatigue, digestive issues, scratchy throat, ear ringing, or fever. They may also gain weight, become irritable, develop blood sugar issues, or feel cold all the time. This progresses into more troubling symptoms, such as brain fog, migraines, heart palpitations, diabetes, memory loss, severe skin rashes, extreme mood swings or chronic fatigue. Over time, they may develop degenerative disease, life-threatening infections or simply collapse, unable to function at all. One day they simply do not show up to work anymore. Their bodies have stopped producing energy and they wither away. For others, cancer or organ failure has taken them.

Those that can no longer afford to pay for their rent and pharmaceuticals are evicted from their homes. Many end up living on the streets in tents, cardboard boxes, and encampments near freeways, struggling to pay for drugs to keep them alive. They are frequently arrested by municipal authorities for living on the streets. The luckier ones seek shelter in the homes of friends and relatives. They stay mostly out of mainstream news and out of the public eye, which does not want to see them. Many people believe these individuals are lazy, unwilling to work, or make very poor lifestyle choices to end up on the streets. You prefer to simply not think about them.

Then there are the wealthy, who seem relatively immune to these problems. They build their houses out of eco materials, filter their air and water, wear clothing from natural fibers, purchase sustainable furniture the working class cannot afford, and take expensive vitamins. They live in gated communities and seldom venture into working class neighborhoods. They are in good health and have the longest lifespan.

You wish you were wealthy because they seem to be the only ones that can reliably bear children. You love kids, but you are part of the working population, which means you are probably infertile. You have not saved enough money to afford the special treatments offered at fertility clinics. Plus, you are not even married. You get on dating websites on occasion because you feel you are supposed to do that. But admittedly you—along with most of your peers—are not interested in having sex at all. For this reason, you worry about meeting someone and not being able to fulfill their romantic expectations.

To distract yourself from these thoughts, you throw yourself into your work. You are a manager at a factory farm, where you oversee vegetable production. Most fruits and vegetables are now pollinated using drones or sometimes hand-pollinated because there are very little insects in the world. Workers and robots remove super-weeds and pests by hand, as many invasive species have evolved to be physically large, extremely stout, and resistant to pesticides.

Your best friend works as a chemist in meat production. Meat is mostly synthesized in test tubes because live animals were getting too sick, making them cost-prohibitive to maintain. Livestock is now a thing of the past. Wild animals are also rare to find but there do seem to be plenty of aggressive species of mosquitos and disease-bearing ticks around. Every city and neighborhood is regularly sprayed for vector control.

Your grandmother tells you that things were not always this way. She claims that when she was a kid, she did not have to take any drugs for her health at all. She remembers swimming in waters that were safe to be in and never had to pay extra for clean drinking water. There were birds, mammals, and fish in the wilderness. Bees, butterflies and hummingbirds would visit her backyard. She also claims that during her own grandmother's lifetime, problems like breast cancer and autism were practically unheard of. She says that men and women were strong and capable of hiking for miles and could do physical work all day without getting fatigued.

You do not really believe her, however. After all, you are told that it has always been like this. You have been taught that industry has made a higher quality of life possible for you, one that was never possible before. That synthesized chemistry is the only way to keep people alive and healthy in nature's hostile world. That before big companies saved the day, people would frequently die of metabolic disorders, cancers, and immunological problems. And that very few human beings on the planet actually had the privilege of reproduction throughout the history of mankind. Indeed, the majority of the human race would not survive the natural world without chemical intervention. It is understood that the pharmaceutical industry has given you the privilege of life; a life you would never enjoy in any other time in human history. You are safe knowing that the chemical industry is making your life possible.

Moreover, every schoolchild of your generation knows that without big agribusiness, nature would never provide enough food for the human race. The collapse of the pollinator populations, mosquito-eating bats, and many marine species—which occurred because of mysterious fungal infections— is proof that nature is untrustworthy when it comes to food security (some people insist that the pollinator population collapsed as a result of toxic pesticides and EMF pollution; but those are just conspiracy theorists). The life-saving role of the chemical industry is taught in books and on the media, and even in grade school.

There are rumors of people who have rejected this system, living in isolated colonies in giant bio-domes in the harshest parts of the desert and wilderness. These bio-domes are reported to be enclosed, reconstructed natural environments with plants, fish, insects, filtered air, water, and soil. They sound like some bizarre science class experiments. Surely the inhabitants of these colonies cannot survive without pharmaceutical drugs for more than a few months. Living in nature like this for a year or more—without scientific intervention—sounds like a religious miracle. These refugees are reported to live without plumbing and electricity, smuggling plants, animals, and supplies from the civilized world.

There are also a number of people among civilized society who believe that chemicals and EMF radiation have caused human dependence on pharmaceuticals in order to live. They carry special wrist watches that warn them when they have exceeded their exposure limits to certain chemicals and EMF pollution. Some have special guide dogs that help them avoid walking into areas treated with pesticides or contaminated with industrial chemicals. These people stay away from certain areas of town and will never enter public buildings. However, they must be delusional because the federal government has told you that your environment is safe.

While this sounds like an outlandish science-fiction story, it is unfortunately the general trajectory of where we are heading as a society. Toxic chemicals that are used freely by industry are major contributors to chronic illness.[3] These toxic substances are so prevalent in our lives that they are virtually unavoidable. What makes them so problematic is that they are similar enough to the molecules found in nature that our body actively engages with them. And yet, they are sufficiently foreign and antagonistic to cause serious problems.

---

[3] EMF radiation is a closely-related environmental problem that leads to similar types types of damage in the human body as toxic chemicals. The literature on EMF radiation is extensive enough to merit its own book. See e.g., Alicja Bortkiewicz, et al., "Mobile phone use and risk for intracranial tumors and salivary gland tumors — A meta-analysis," *International Journal of Occupational Medicine & Environmental Health* 30(1), 27–43 (2017); Houston, BJ, et al., "The effects of radiofrequency electromagnetic radiation on sperm function," *Reproduction* 152(6), R263-R276 (Dec. 2016); Cindy Sage and David O. Carpenter (eds.) "Bioinitiative 2012: A Rationale for Biologically-based Exposure Standards for Low-Intensity Electromagnetic Radiation" BioInitiative Working Group (Dec. 2012); Carl Blackman, "Cell phone radiation: Evidence from ELF and RF studies supporting more inclusive risk identification and assessment," *Pathophysiology* 16(2-3), 205-216 (Aug 2009); Karl Hecht, et al., "Health Implications of Long-Term Exposure to Electrosmog," Competence Initiative for the Protection of Humanity, the Environment and Democracy, Brochure 6 (2017), available at https://www.emfanalysis.com/wp-content/uploads/2017/01/German-Report-on-878-Russian-EMF-Health-Studies.pdf.

The scientific literature regarding the hazards of toxic chemicals goes back nearly 60 years. Mainstream society, however, is mostly unaware because the voices of the scientific community have been suppressed, ignored, and actively attacked by industry at every turn. Our government agencies and public institutions—the ones we most trust—have long ago been recruited to represent industry's best interests and keep us in the dark. The amount of energy and resources that have gone into suppressing our toxic reality is astronomical and has stunted our social, technological, and economic progress. It is costing us our health and putting our ecological survival at risk.

Indeed, our current economic model is based on the historical paradigm of resource destruction called colonialism—we invade new territories, destroy the local environment, build man-made infrastructure to replace what we have destroyed, and make the locals dependent on this man-made infrastructure. We then sell essentials back to people who would otherwise live from the land, and force them to participate in a colonial economy. Once we exhaust the local natural resources, we confiscate the profits and move on.[4]

The inevitable next target of colonization is the human body itself. Our economic activities are polluting and poisoning each of us.[5] The sicker we get from chemical pollution, the more reliant we become on industry to provide us with biotech and pharmaceutical solutions to maintain our good health. Chemicals are sold to us that make us sick, and chemicals are sold to us to make us feel better. At some point, we will be in such poor health that we will become fully dependent on the chemical industry to survive. If we continue this path, we will effectively become slaves to the status quo—working to pay our medical bills so that we can stay functional enough to continue to work to pay our medical bills.

Indeed, some of us have already reached this turning point. People who suffer from chronic illnesses such as cancer and chronic fatigue—two diseases catalyzed by toxic chemicals—are already desperately searching for drug cures so that they can continue with their normal lives, show up for work, and bring in an income to survive. The drugs for diseases tied to toxic chemical exposure are very expensive—so expensive that many people can no longer afford them. Some have died from their inability to pay. Diabetes is a prime example:

---

[4] Raj Patel & Jason W. Moore, *A History of the World in Seven Cheap Things* (University of California Press 2017); see also Vandana Shiva, *Oneness v. The 1%*, 24-25 (Spinifex Press 2018).

[5] Theo Colborn, et al., *Our Stolen Future* (Penguin Group 1996).

> The price [for diabetics] is so high that people are doing desperate things to get by, like using expired insulin, relying on crowdfunding to pay their bills, or taking less insulin than they need in an effort to ration their supplies.[6]

While the current trajectory is a profitable one for corporations, it is also a zero sum game for society. We will never thrive if we are physically and economically dependent on man-made chemicals. Rather, we will be taken advantage of by the chemical industry and those who profit from its growth. We will be more beholden to their interests and even less capable of asserting our needs for a life-supporting environment, economy, and political system. It is thus imperative for all of us that we turn this ship around.

The only way to get off this path is to change our status quo. The only way to change our status quo is to change our individual mindsets about how we should operate in this world. We may think we are helpless as to what happens in our companies and economies, but in reality each of us—in everything we chose to say and do—either contribute to the problem or help solve it. Each of us holds the responsibility to make decisions from moment to moment that speak our truth and support a sustainable world. If we continue to shirk that responsibility out of fear-based self-interest, deny what is happening and suppress how we feel about the status quo, we will find ourselves enslaved to a reality that we have created.

---

[6] "High Insulin Costs are Killing Americans," Rightcare Alliance Website (last visited Jan. 18, 2019), available at https://rightcarealliance.org/actions/insulin; see also Chapter 16 for further discussion regarding diabetes.

CHAPTER 2: OUR CHEMISTRY

Chemistry is essential to all life on earth. Indeed, it is a combination of radiation and chemistry that caused life to form on this planet. The chemical environment found within each of our bodies is a complex orchestra that has evolved in the natural world over an unfathomable amount of time. We developed into increasingly sophisticated creatures through the manipulation of our own internal body chemistry over countless generations. And chemistry continues to facilitate every single life-supporting function—no matter how small—within every living being on earth. This includes the bacteria, fungi, plants, animals, and us.

Virtually nothing can happen inside any life form—including people—without a chemical being formed or released to catalyze the action. Human growth, metabolism, body temperature, organ function, immune system, thoughts, emotions, movement and reproduction are wholly dependent upon our internal body chemistry. Our genes and DNA lie dormant without the influence of chemistry. Our bodies would be dead without our natural chemistry. Indeed, we run on endless chains of sophisticated chemical reactions.

Biological chemistry has evolved over millions of years, and continues to evolve over time in every living species. New chemical reactions arise to help each species adapt to its environment and resist predators. This is how we ended up with useful things such as photosynthesis, cell division, reproduction, respiration, thoughts, emotions, and countless of other life-giving activities we commonly take for granted.

This life-giving chemistry does not evolve in a vacuum. Plants, animals and micro-organisms co-exist and-evolve in an intricate dance. For example, the bacteria in our guts evolve their chemistry over time to help safely digest foods that would otherwise be toxic; this guarantees a home and a host for them. Plants develop new toxic chemicals in order to be less desirable to those who want to eat them. They also create useful chemicals that encourage other species to pollinate them. Species evolve in their ever-changing environment to adapt and thrive. It is a beautiful and intricate web of life created by nature.

Each species can alter itself within a generation—meaning that you and I can develop new chemistries to adapt to our environment.[7] We may move from a warm climate to a cold one and find it takes some years to physically adapt to winter; but it ultimately can be done. We may be resistant to chicken pox in adulthood after being ill with the disease during childhood. These types of adaptations take place within our lifetime and are dictated by epigenetics, or processes that do not require a change of the genetic code. We already have the mechanisms within our existing set of genes to make these types of physical changes occur.

However, there are many evolutionary processes that take generations to develop because they do require our genetic code to change. For example, we will not grow flippers simply because we go swimming every day. We will not learn to convert sunlight into energy through photosynthesis the way that plants do just by moving to a sunnier climate. Our genes do not have the programming to make this happen. As a result, these changes are too large to take place within one lifetime. Such changes could theoretically happen over the lifetime of the planet, which is long enough for our genes to change.

Whether we successfully adapt to our environment—within the course of one or multiple lifetimes—depends on our genetic starting place and how much time we have. Dinosaurs, for example, were unable to adapt to certain abrupt changes in the climate millions of years ago and became extinct. The changes came too quickly for them to adapt epigenetically or over multiple lifetimes. Native Americans died from disease in massive numbers when Europeans first moved to North America because they could not adapt quickly enough to illnesses that Europeans had grown accustomed to over many centuries. All of us can adapt only so fast to changing conditions. If changes in our environment exceed our physical capacity to adapt, we will not thrive and we may not even survive very long.

Rapid change is required in modern life. Over the last 80–100 years, we have unleashed an unprecedented synthetic chemical cocktail onto our

---

[7] US National Library, "Genetics Home Reference: What is Epigenetics?" National Institutes of Health (March 3, 2020).

economies, communities, and ecosystems. These chemicals have drastically altered our ecosystems and internal body chemistry.[8] In approximately eighty years' time (roughly since the rise of the chemical industry in World War II), we have introduced well over 100,000 synthetic chemicals *en masse* into our lives.[9] Trillions of pounds of synthetic chemicals are put into commerce annually and can be found in personal care items, cleaning products, clothes, food, water and the built environment.[10] We have been exposed to some of these chemicals only in the last few decades. Others went out of popular circulation well over forty years ago but are still found in, and affect, every living being on the planet.[11]

Scientists have been documenting the presence, health effects and the ways in which we get exposed to these toxic chemicals. There are countless research papers and scientific journals about chemical toxicity and human or environmental health. Yet we rarely hear about this research in mainstream media because the issue is not popular with certain sectors. Synthesized chemicals have become a multi-trillion dollar industry and there is a lot of profit at stake for the corporations who use and manufacture them.

By the time Rachel Carson wrote *Silent Spring* in the 1960s, we knew that human beings had widely introduced synthesized chemicals into the environment that were disrupting our genes, interfering with our metabolism, causing cancer, inducing birth defects and extinguishing life.[12] Three decades later, when Theo Colborn wrote "Our Stolen Future," we knew that we were introducing synthetic chemicals into the world that are disrupting our growth and development and that their effects extended to future generations.[13] Another thirty years have passed and we are starting to see to what extent these synthesized chemicals are disrupting global health and well-being.

---

[8] Theo Colborn, et al., *Our Stolen Future* (Penguin Group 1996).

[9] US EPA News Release, "EPA Releases First Major Update to Chemicals List in 40 Years," (February 19, 2019), available at https://www.epa.gov/newsreleases/epa-releases-first-major-update-chemicals-list-40-years. This registry excludes certain industries not regulated from reporting and excludes chemicals used at lower volumes. While not all of these chemicals may be in active use today, it is still highly likely that the number underestimates their true extent.

[10] United Nations Environment Programme, "Global Chemicals Outlook II" (2019); see also Joanna Malaczynski, "Chemicals to Avoid in Consumer Products — Monographs for Purchasers," DESi Potential (April 7, 2020).

[11] US Centers for Disease Control, "Fourth National Report on Human Exposure to Environmental Chemicals" (2019).

[12] Rachel Carson, *Silent Spring* (The Riverside Press 1962).

[13] Theo Colborn, et al., *Our Stolen Future* (Penguin Group 1996).

> [S]ynthetic chemicals have become so pervasive in the environ-
> ment and in our bodies that it is no longer possible to define a normal,
> unaltered human physiology.
>
> —Theo Colborn, 1996[14]

Human growth, development, and basic physiology have been under-
mined by toxic chemicals, leading to a proliferation of illnesses such as
cancer, diabetes, heart disease, kidney failure, chronic fatigue, autism and
a whole host of other acute and chronic illnesses (Chapters 13–20). We are
becoming sicker, fatter and more prone to disease with each generation. Even
our pets are suffering with the same diseases—thyroid problems, kidney
failure, allergies and cancer are all now common in domesticated animals.
And for the first time in recent history, we have a decrease in human life
expectancy.[15] Longevity and good health peaked around the Baby Boomer
generation (born before 1965). Those who have come after are not faring so
well. The youngest generations are the hardest hit, facing an accumulation
of our chemical legacy.

> Millennials [today's young adults born between 1980 and 1996] are
> seeing their health decline faster than the previous generation as they
> age. This extends to both physical health conditions, such as hyper-
> tension and high cholesterol, and behavioral health conditions, such
> as major depression and hyperactivity. Without intervention, millen-
> nials could feasibly see mortality rates climb up by more than 40%
> compared to [Generation X—those born between 1965 and 1980] at
> the same age.[16]

Little has been done by policymakers and regulatory agencies to intervene in this growing epidemic. This is because governments have largely
been overtaken by corporate interests. Moreover, the human health organizations that are commonly believed to represent our best interests are not
able to do so. Rather, they often get steamrolled by chemical industry bullies
and so they maintain scientifically unsound or outdated assertions in public-
facing literature.

At the same time, we are consistently told that our health is a function
of our genes, stress levels and similar lifestyle factors. Those in any position

---

[14] Theo Colborn, et al., *Our Stolen Future*, 240 (Penguin Group 1996).

[15] Swanson, N.L., Leu, A., Abrahamson, J. & Wallet, B. "Genetically engineered crops,
glyphosate and the deterioration of health in the United States of America," *Journal of
Organic Systems* 9(2), 6–37 (January 2014).

[16] Moody's Analytics, "The Economic Consequences of Millennial Health," Blue Cross Blue
Shield (Nov. 6, 2019), available at https://www.bcbs.com/sites/default/files/file-attach-
ments/health-of-america-report/HOA-Moodys-Millennial-10-30.pdf.

of authority do not talk to us about the major role that synthetic chemicals have to play. Either they do not know about this themselves or it is politically convenient for them to gloss over the facts. As a result, we are willing to blame ourselves for the rise in chronic illness.

Not everyone will get sick at once, to the same degree and with the same disease. Our health outcomes are a complex equation dependent upon our unique body chemistry, other environmental stressors and the circumstances of our exposure. And because we all experience environmental pollution differently, it may be easy for some of us to believe that toxic chemicals do not have anything to do with our health problems. After all, we do not want to believe that we have been duped by society or that the institutions we trust are not so trustworthy after all.

CHAPTER 3: THE STATUS QUO

> What is more important for the long-term health of a nation than the health of its citizens? A nation grows weak as its people fall to disease. [We have the] right to be born into a healthy environment, an environment that will not cause chronic childhood and adult disease.[17]

We can be exposed to toxic chemicals at seemingly trivial quantities and become very ill—under the right circumstances, parts per trillion is enough to alter our growth, development and long-term well-being.[18] How much is that? Very little. One drop of liquid in a railroad car tank is the equivalent of a part per billion. A grain of sugar in an Olympic size swimming pool is the equivalent of a part per trillion.[19] Humans can detect odors at parts per billion.[20] Dogs are able to detect smells at parts per trillion.[21] And although they come in pills and liquids that are mostly filler, pharmaceutical drugs

---

[17] "The 28th Amendment Project," Naviaux Lab (last visited February 19, 2020).

[18] Agency for Toxic Substances and Disease Registry, "Health Effects of Chemical Exposure," (July 15, 2019), available at https://www.atsdr.cdc.gov/emes/public/docs/Health%20Effects%20of%20Chemical%20Exposure%20FS.pdf; US EPA, "Drinking Water Requirements for States and Public Water Systems—Chemical Contaminant Rules," available at https://www.epa.gov/dwreginfo/chemical-contaminant-rules#rule-summary (last visited July 15, 2019); US EPA "Lifetime Health Advisories and Health Effects Support Documents for Perfluorooctanoic Acid and Perfluorooctane Sulfonate," Federal Register 81(101) (May 25, 2016). See also Chapters 15-17 on hormone disruptors in this book.

[19] AE Marchewski, "How Much is A Part Per Million?" Michigan State University Extension Service, Ag Facts, Extension Bulletin E-1641 (August 1982).

[20] Lee Sela and Noam Sobel, "Human olfaction: a constant state of change-blindness," *Experimental Brain Research* 205(1), 13-229 (August 2010).

[21] Peter Tyson, "Dogs' Dazzling Sense of Smell," *PBS NOVA* (Oct. 3, 2012).

---

that have life-altering and life-saving effects are often times dosed to us in parts per billion.[22]

Moreover, each of us can have dozens of toxic chemical exposures in a given day.[23] We may wake up and apply a handful of personal care products to our bodies with a number of questionable chemicals in them.[24] We put on synthetic and chemically treated clothes, exposing us to chemicals through our skin.[25] We eat breakfast tainted with pesticides and chemicals found in food packaging.[26] We get exposed to chemical vapors from the cleaning products and building materials in our workplaces.[27] Even if we are only exposed to small quantities of any given chemical, it all adds up to a lot. It is a bit like going to the grocery store and being surprised that our grocery bill exceeded $200 when almost all of the items we bought cost only a few dollars. Toxic chemicals can also add up—and our bodies do feel the burden.

To make matters worse, any given consumer product can be made up of hundreds of chemicals. A backpack, for example, contains a whole universe of different chemicals in each of its components. There is the exterior shell and the inner padding. There are the zippers and strings, the adjustable shoulder straps and the water bottle holder. Each of these components is made up of different materials. These may include metals and coatings in the zipper or dyes and synthetic fabrics in the shell. And each of these materials

[22] E.g., a birth control pill containing 0.35 mg of active hormone is enough to prevent pregnancy in a 132-pound (60 kg) woman. National Institutes of Health, US National Library of Medicine, "Ortho-Micronor Norethindrone Tablet" *DailyMed Label* (updated May 23, 2019). The EpiPen, used to prevent life-threatening allergic reactions, is administered at a dose of 0.3 mg for children weighing 66 pounds (30 kg) or more. National Institutes of Health, US National Library of Medicine, "EpiPen, Epinephrine Injection" *DailyMed Label* (updated January 15, 2020).

[23] See e.g., the State of Washington Department of Ecology Children's Safe Product Act, "Manufacturer Reporting," available at https://ecology.wa.gov/Waste-Toxics/Reducing-toxic-chemicals/Childrens-Safe-Products-Act (last visited April 29, 2020) (self-reported data by industry of priority chemicals present in products marketed toward children, such as shoes, clothes, furniture, toys, etc.).

[24] See e.g., "Teen Girls' Body Burden of Hormone-Altering Cosmetics Chemicals: Detailed Finding," Environmental Working Group (Sept. 24, 2008), available at https://www.ewg.org/research/teen-girls-body-burden-hormone-altering-cosmetics-chemicals/detailed-findings and Lauren Zanolli, "Pretty Hurts: are chemicals in beauty products making us ill?" *The Guardian* (May 23, 2019).

[25] Greenpeace International, "Toxic Threads: The Big Fashion Stitch-Up" (2012).

[26] Carey Gillam, *Whitewash*, ch. 4 (Island Press 2017); see also Chapter 14 in this book.

[27] PPT Presentation by David Kunz, "Addressing Chemicals of Concern" Oregon Department of Environmental Quality, presented at the 2014 Pacific Northwest Pollution Prevention Resource Center Pollution Prevention Roundtable, available at https://pprc.org/wp-content/uploads/2014/11/P2-Roundtable-2014.pdf; and Agency for Toxic Substances and Disease Registry, "Health Effects of Chemical Exposure," available at https://www.atsdr.cdc.gov/emes/public/docs/Health%20Effects%20of%20Chemical%20Exposure%20FS.pdf (last visited July 15, 2019).

is comprised of its own universe of chemicals. The zipper may be a metal alloy coated with a rust-proof coating. The shell may be a synthetic fabric with its own dyes and coatings.[28]

About 2.3 billion metric tons of synthetic chemicals are used annually today.[29] That means more synthetic chemicals are produced on this planet in one year than there are people in body weight on the entirety of this globe. These are synthetic chemical combinations that did not exist in nature, say, 100 years ago. Life on this planet has never known these chemistries before. Think of how much extra work that is for our bodies—and for every living being on this planet. We have to process these chemicals through our system upon exposure.

How many different synthetic chemicals are there? Probably hundreds of thousands. Over eighty-thousand synthetic chemicals have been registered with the US EPA and many are actively used in US manufacturing and consumer products.[30] This eighty-thousand does not include pharmaceuticals, food additives, and other chemicals that need not be reported to the EPA.

The chemicals that are most problematic are those that are persistent, bio-accumulative, and toxic (PBTs). These are chemicals that are very resistant to breaking down—it may take years, decades, and sometimes centuries or more for them to do so. They also build up in our bodies and the food chain. The more we are exposed to them, the more they build up in our bodies. The more we eat plants and animals that have been exposed to them, the more they accumulate in our bodies. They are also toxic. They can poison us quickly or slowly, depending on the context and level of exposure.

Such chemicals are actively used throughout consumer products and our economy. They are found in our personal care products, cleaning products, plastics, electronics, pesticides, clothes, furniture, and even in our homes and offices. The most common PBTs throughout history have been heavy

---

[28] Joanna Malaczynski, "Chemicals of Concern in Consumer Products—Where Are They?," DESi Potential (March 20, 2020); see also the State of Washington Department of Ecology Children's Safe Product Act, "Manufacturer Reporting," available at https://ecology.wa.gov/Waste-Toxics/Reducing-toxic-chemicals/Childrens-Safe-Products-Act (last visited April 29, 2020 (self-reported data by industry of priority chemicals present in products marketed toward children, such as shoes, clothes, furniture, toys, etc.).

[29] Based on chemical industry capacity figures. See United Nations Environment Programme, "Global Chemicals Outlook II" 27, SBN No: 978-92-807-3745-5 (2019).

[30] US EPA News Release, "EPA Releases First Major Update to Chemicals List in 40 Years," (Feb. 19, 2019), available at https://www.epa.gov/newsreleases/epa-releases-first-major-update-chemicals-list-40-years. This registry excludes certain industries not regulated from reporting and excludes chemicals used at lower volumes. While not all of these chemicals may be in active use today, it is still highly likely that the number underestimates their true extent.

metals—lead, mercury, cadmium, arsenic, etc. Despite our recognition for hundreds of years of their toxicity, we continue to use these substances liberally and struggle to manage their use in a safe manner. Worse yet, we have added many more persistent, bio-accumulative and toxic chemicals to our world with synthesized chemistry. In the last century, we unleashed DDT, PCBs, dioxins, many biocides, PFAS, and others. These chemicals are found in the blood streams of virtually every living creature—including human beings—around the globe.[31]

Scientists have only studied a small subset of these chemicals. Nevertheless, they can frequently predict that similar substances will be problematic.[32] For example, BPA is a known toxic chemical and hormone disruptor (see Chapters 15 and 23) used in plastics and packaging. It has been substituted with another chemical known as BPS in some products. As it turns out, PBA and BPS are both bisphenol compounds. Because they are chemically very similar, they can be predicted to have similar hormone disrupting effects. Research is now confirming that this is the case.[33]

Unfortunately, just because a new chemical is structurally different from a known toxic chemical does not mean that it is safe for us. We use two-dimensional models to predict whether substances are chemically similar. In real life, however, chemicals are complex, organic 3-D entities.[34] They cannot be simplified into two-dimensional models for purposes of predicting their impact on our bodies. As a result, two chemicals that look very different from each other can still both be highly problematic.[35]

You probably are not aware of the existence of all these chemicals because companies are generally not required to disclose them to you. And many companies are not aware of all of the chemicals that are in their own products. They themselves buy materials and components from someone else. And usually there is more than one third-party involved; there may be

---

[31] US Centers for Disease Control, "Fourth National Report on Human Exposure to Environmental Chemicals" (2019) and previous reports dating back to US Centers for Disease Control, "National Report on Human Exposure to Environmental Chemicals (2001). See also European Union, "Endocrine Disruptors: from Scientific Evidence to Human Health Protection," a study commissioned by the PETI Committee of the European Parliament PE 608.866 (March 2019).

[32] M.D. Barratt, "Prediction of toxicity from chemical structure," *Cell Biology and Toxicology* 16(1), 1-13 (Feb. 2000).

[33] Soria Eladak, et al., "A new chapter in the bisphenol A story: bisphenol S and bisphenol F are not safe alternatives to this compound," *Fertility & Sterility* 103(1), 11-21 (January 2015).

[34] See "Help Me Understand Genetics: How Do Genes Work" In *Genetics Home Reference*, U.S. National Library of Medicine (January 2020).

[35] Theo Colborn, et al., *Our Stolen Future*, 75, 84 (Penguin Group 1997).

a whole host of manufacturers making various chemicals, materials, and pieces that make up the final product.[36]

There have been efforts in recent years by some brands to identify the chemicals in their products. And some brands have gone so far as to request that certain chemicals not be used by their suppliers. Many toxic substances continue to be used even by our most progressive brands, however, because the perceived economic benefits of using the chemicals—absent regulation and environmental accountability—outweigh the financial risks of doing so.

Unfortunately, we generally avoid holding corporations accountable for making and using toxic chemicals. As a result, industry is perfectly content with the status quo and works hard to protect it—even if it results in the widespread exposure of society and our environment to problematic substances. For some industries, toxic chemicals are so integral to what they do that their entire business models depend on them. Thus they go to great lengths to cover up their chemical toxicity in order to avoid having to change.

---

[36] Joanna Malaczynski, "Chemicals of Concern in Consumer Products—Where Are They?," DESi Potential (March 20, 2020).

CHAPTER 4: COLLUSION, COMPLICITY & CORRUPTION

The evidence is overwhelming: major corporations have enlisted our regulatory agencies in covering up the health effects of toxic substances. We know this as a result of countless internal company documents that have been amassed through years of litigation, investigation, and Freedom of Information Act requests.[37] These documents come from some of the largest companies in the world, including Monsanto, DuPont, 3M, Alcoa, Dow, Phillip Morris, Chevron, Coca-Cola, and many others. In all the documents and data, there is a clear trend across many industries that companies are covering up the hazards of their products and successfully manipulating the regulatory agencies to convince the public that we are not in fact poisoning ourselves and our environment.

The poster child for the problem is the US Toxic Substances and Control Act (TSCA). The law was off to a bad start even before its passage in 1976. It grandfathered the existing 62,000 synthetic chemicals already in commercial use.[38] New toxic chemicals were not regulated effectively either. Under TSCA, industry does not have to prove that the chemicals they seek to manufacture, import, sell or use are safe for human beings or the environment.

---

[37] See Jonathan R. Latham, et al., "Poison Papers, The Bioscience Resource Project and The Center for Media and Democracy" (2017), available at https://www.poisonpapers.org), Chowkwanyun, Merlin, et al., Toxic Docs 1.0, Columbia University and City University of New York (2018), available at http://www.toxicdocs.org.

[38] See e.g., Presentation by Marcus Aguilar, "Toxic Substances Control Act," US EPA Region 9 (May 4, 2017), available at https://www.epa.gov/sites/production/files/2017-05/documents/toxic_substances_conrol_act_may_2017.pdf and "Off the Books II: More Secret Chemicals: TSCA," Environmental Working Group (May 9, 2016).

Rather, the burden is on the EPA to determine that a chemical is unsafe.[39] That burden is so high—and the politics are so hot—that this has almost never happened in the history of the US EPA. As a result, the agency has managed to regulate only a small a handful of chemicals in its 40 or so years of existence—even of the many chemicals it officially does consider to be toxic.[40]

In an attempt to improve the EPA's performance with TSCA, the law was updated in 2017.[41] While some small tweaks were made, very little has changed. TSCA now requires the EPA to review chemicals on a "regular" basis to speed the process up. However, common estimates predict that it would take the agency centuries to review its existing backlog of chemicals in addition to the hundreds—if not thousands—of chemicals that enter the market annually.[42]

At the same time, more than 2,500 chemicals used in commerce are recognized as being sufficiently toxic to human and environmental health by regulatory agencies across the globe that they have become potential priorities for regulation in the State of California.[43]

However, no one is effectively regulating them in California, either.[44] And the European Environment Agency, a reputably more effective agency than the US EPA, has stated that over 60% of all of the chemicals consumed in the European Union are hazardous to human health.[45] And yet not even the EU has been able to successfully manage their use.

---

[39] Frank R. Lautenberg Chemical Safety for the 21st Century Act, Toxic Substances Control—Prioritization, risk evaluation, and regulation of chemical substances and mixtures, 15 USC §2605 (2016).

[40] Sheldon Krimsky, "The unsteady state and inertia of chemical regulation under the US Toxic Substances Control Act," *PLOS Biology* 15(12) (Dec 18, 2017).

[41] Frank R. Lautenberg Chemical Safety for the 21st Century Act 15 USC § 2601 et seq. (2016).

[42] Mark Scialla, "It could take centuries for EPA to test all the unregulated chemicals under a new landmark bill," *PBS Science* (June 22, 2016). See also Sheldon Krimsky, "The unsteady state and inertia of chemical regulation under the US Toxic Substances Control Act," *PLOS Biology* 15(12) (Dec 18, 2017).

[43] California DTSC "Safer Consumer Products Candidate Chemicals List," available at https://dtsc.ca.gov/scp/candidate-chemicals-list/#! (last visited May 29, 2019) (this list is comprised of chemicals that are priorities for regulatory agencies within the US and abroad).

[44] Some years ago, I established a company called EcoValuate to help companies meet emerging compliance requirements under California's Safer Consumer Product Regulations. Soon after the regulations came into effect, they politically imploded upon themselves. My anticipated clients went from needing to develop greener products right away to not worrying about the issue at all.

[45] United Nations Environment Programme, "Global Chemicals Outlook II" ISBN No: 978-92-807-3745-5 (2019).

Why are US regulatory systems so ineffective? Because regulatory agencies are highly vulnerable to political pressures. And industry has gotten very good at applying that pressure. If an agency gets involved in investigations or activities that are politically unfavorable to those with deep pockets, they will get a threatening phone call.[46] If individual staff members document safety issues with hazardous chemicals, they are asked to revise their reports, treated badly by their superiors, and/or transferred to another department.[47] This makes for a very uncomfortable environment.

I have interacted with a number of agency staff. Many of them are intelligent and well-intentioned individuals who care about our health. However, they are under continuous pressure by industry, lobbyists, and politics to back off and not make any waves. And everything they say and do is under constant scrutiny. The unfortunate consequence is that they adapt to a culture of passivity in order to avoid undesired attention. Some of them also lose their sense of priority in order to advance their careers. In an environment that socially looks down upon you for calling attention to problems and politically rewards you for letting things slide, it is tempting to do so.

Those who butt heads with the system have a very difficult time. An example of this is EPA whistleblower Cate Jenkins. In the early 1990s, Jenkins brought to light how Monsanto actively covered up the health dangers of dioxin, a highly toxic chemical that is present in Agent Orange and some of Monsanto's other products. This came at a time when Vietnam veterans were still fighting for their right to obtain compensation for the horrible diseases they were developing as a result of their exposure to Agent Orange.[48] Vietnam veterans were denied medical benefits because Monsanto claimed dioxins were safe.[49] They were also commonly told they were crazy by society and the medical establishment for claiming a problem existed.

EPA whistleblower Cate Jenkins exposed how Monsanto doctored research studies about dioxin and even covered up the presence of dioxin in consumer products—such as Lysol disinfectant—in order to protect its

---

[46] A former head of an investigative branch of the federal government who was also one of my law professors admitted in my law school class sessions of having received such phone calls.

[47] See e.g., Shiv Chopra, *Corrupt to the Core: Memoirs of a Health Canada Whistleblower* (KOS Publishing 2009).

[48] Rachel's Hazardous Waste News #212, "Report Links Herbicide Exposure to Illness Among Vietnam Vets," Environmental Research Foundation (December 19, 1990), available at toxicdocs.org.

[49] Interview of William Sanjour (former EPA employee) and Gerson Smoger (attorney representing Vietnam veterans) in the documentary by Marie-Monique Robin entitled, *The World According to Monsanto* (2008).

market share.[50] Jenkins also expressed concern to her superiors that the EPA had relied on Monsanto's fraudulent research, allowing dioxins to continue on the market.[51] As repayment for her efforts, Jenkins was actively harassed both by the EPA and by Monsanto for challenging the Monsanto empire.[52] The EPA purported to investigate Monsanto's inappropriate conduct but took little action toward that goal and ultimately dropped its efforts.

> The fact is—that there was no investigation of Monsanto. It did not exist. Nobody investigated the [dioxin] studies. Nobody; period. What [the EPA] investigated was Cate Jenkins, the whistleblower. They made her life a hell. They harassed her, they changed her jobs, they persecuted the poor woman.
>
> —Former EPA employee William Sanjour[53]

The EPA withdrew all of Jenkins' existing work assignments and then gave her clerical work. This caused Jenkins to file a harassment complaint with the Department of Labor. During the subsequent Department of Labor investigation, one of Jenkins' supervisors was quoted as saying, "Cate appears to be on a mission to eliminate dioxins wherever they may be... Cate Jenkins has a very intelligent mind and it's a shame it can't be focused differently."[54] What is interesting about this statement is how dismissive it is of Jenkins and how it seeks to shame her for doing her job. Jenkins was not overstepping her boundaries; dioxins are extremely toxic and controlling their production should have been the EPA's top priority.

At the same time, Monsanto was busy harassing Jenkins as well. They accused her in official letters to the EPA of acting outside the scope of her authority, asserted she did not have the scientific know-how to do her job, and assured federal authorities that Monsanto's scientific studies were

---

[50] Memo from Cate Jenkins, USEPA Regulatory Development Branch to John West, USEPA Office of Criminal Investigations, re "Criminal Investigation of Monsanto Corporation: Cover-up of Dioxin Contamination in Products / Falsification of Dioxin Health Studies," dated November 15, 1990, available at toxdocs.org.

[51] Memo from Cate Jenkins, US EPA Characterization and Assessment Division to Raymond Loehr, US EPA Office of the Administrator, "Newly Revealed Fraud by Monsanto in an Epidemiological Study Used by EPA to Assess Human Health Effects from Dioxins," dated February 23, 1990, available at toxdocs.org.

[52] Memo from William Sanjour, US EPA Policy Analyst to David Bussard, US EPA Characterization and Assessment Division Director, entitled, "The Monsanto Investigation," dated July 20, 1994, available at toxicdocs.org.

[53] Interview of former EPA employee William Sanjour in the documentary by Marie-Monique Robin entitled, *The World According to Monsanto* (2008).

[54] Memo from William Sanjour, US EPA Policy Analyst to David Bussard, US EPA Characterization and Assessment Division Director, entitled, "The Monsanto Investigation" at p. 20, dated July 20, 1994, available at toxicdocs.org.

accurate.[55] In reality, internal corporate documents showed that Monsanto had known about the dangers of dioxins for decades. Monsanto had even collated studies on dioxins going back to the 1940s, identifying their terrible health effects, including birth defects, liver disorders, a cyst-like acne known as chloracne, cancer, miscarriages, and other serious health conditions.[56]

In Canada, things have not been any better. The lead environmental agency, Health Canada, claims that it is committed to "helping Canadians maintain and improve their health." In private, the agency has gone so far as to describe industry as its client; the public was not included in their "client" definition at all.[57] Regulatory staff members became so frustrated with how things were going that they wrote a formal letter to their union complaining of corruption mobilized by industry lobbies, stating as follows:

> [We have been facing] many years of pressure and harassment to pass various products and processes of questionable safety....[Upper management] frequently order us to alter our decisions and recommendations to please the pharmaceutical and other industries on such critical issues as carcinogenicity [i.e. propensity to cause cancer], antibiotic resistance and other harmful effects on people.[58]

EPA whistleblower EG Vallianatos notes that the political pressure on staff to water down environmental regulation comes from all directions and crosses party lines.[59] It does not matter whether the current administration is republican or democratic or of any other ideology. Politics protect corporate interests over public interests. And the political influence of industry has grown exponentially over time with the proliferation of super PACs, corrupt campaign financing laws, Supreme Court decisions favoring corporate interests, limits on plaintiff's lawsuits, and growing oligopoly power.

> The fact of the matter is that the chemical industry has a tremendous amount of power, influence and resources and they have deployed that for decades against the interests of the average person out there. They've done it with lobbyists, they've done it with campaign contri-

---

[55] Letter from James Collins, Monsanto Epidemiology Director to Marilyn Fingerhut, Chief of NIOSH Industrywide Studies Branch, dated June 1, 1990, available at toxicdocs.org.

[56] Attachment to Letter from Allan M. Ford, Monsanto, to Myron Weinberg, Weinberg Consulting Group, dated February 24, 1984, available at toxicdocs.org.

[57] Shiv Chopra, *Corrupt to the Core: Memoirs of a Health Canada Whistleblower*, 70-71 (KOS Publishing 2009).

[58] Shiv Chopra, *Corrupt to the Core: Memoirs of a Health Canada Whistleblower*, 254-246 (KOS Publishing 2009), describing and quoting from official 2001 letter sent by Shiv Chopra and five other staff members.

[59] E.G. Vallianatos with M. Jenkins, *Poison Spring*, ix (Bloomsbury Press 2014).

butions, they've done it by buying studies that then masquerade as science and this continues to go on.

—Congressman John Sarbanes[60]

---

[60] Testimony of Congressman John Sarbanes in front of the US House of Representatives Committee on Oversight & Reform (116th Congress), during the "The Devil They Knew: PFAS Contamination and the Need for Corporate Accountability, Part II" Hearing (Sept. 10, 2019).

Chapter 5: Rubber Stamping

While industry becomes increasingly wealthy and powerful from year to year, the regulatory agencies become weaker. A budget cut (or failure to renew a budget allocation) can significantly obliterate the resources an agency has to investigate and enforce regulations that are already in effect. Budget cuts regularly plague our environmental agencies.[61] For this reason, it is increasingly common for our regulatory staff to rely on industry to supply critical health and safety information. Indeed, the EPA relies almost exclusively upon laboratories paid for by industry for its chemical toxicity data.[62]

And most of the toxicity studies submitted by industry tend to be biased.[63] The staff members at these labs are under enormous pressure to please the party holding the purse strings.[64] The client—meaning, the chemical industry—wants favorable results and these labs can be eager to deliver them. Having worked as a consultant to both industry and government, I can attest that there is a strong culture among service firms to make the client happy—even if it means sweeping troublesome data under the rug. The perception is that any inconvenience to the client might mean losing their business. As a result, pointing out problems is usually frowned upon in these circles. In certain instances, problems are not only downplayed—they are actively covered up.

---

[61] Jennifer Beth Sass & Mae Wu, "Budget Cuts to the EPA will Reduce Government Data on Pollutants, and Increase Reliance on Industry Data," *International Journal of Occupational & Environmental Health* 13:2 (2007).

[62] E.G. Vallianatos with M. Jenkins, *Poison Spring,* 3 (Bloomsbury Press 2014).

[63] David Michaels, "Science for Sale," *Boston Review* (January 28, 2020).

[64] E.G. Vallianatos with M. Jenkins, *Poison Spring,* ch. 7 (Bloomsbury Press 2014).

This is a sufficient weakness in the system to allow toxic chemicals to pass regulatory scrutiny. Our federal government has relied upon thousands of fraudulent tests from industry labs to bless many chemicals used by consumers today.[65] In one famous case, the EPA found out about fraudulent data implicating 4,500 lab tests, 80% of which the agency suspected were invalid.[66] What did the agency do? It held a meeting with industry representatives to announce it would sweep the problem under the rug:

> At one time, we discussed what the options of the agency were. The most extreme would have been to purge our files of all [implicated lab] data and either impose a data requirement on the registrants or initiate a more extreme action....We determined that was neither in EPA's interest or the public interest or the registrants' interest....[67]

The agency went on to say that at least two other countries—including Canada—had also relied on this fraudulent data. These countries were initially prepared to pull the chemicals at issue off of the market. But after consulting with the EPA, they were persuaded to let things slide and defer to the US regulators on what to do.[68]

What a windfall for industry. Not only did they get away with fraud for years, but they managed to keep all the underlying chemicals in the marketplace, even after their fraudulent toxicity data had been revealed. The EPA absolved itself of the need to review this data, which would have given them a headache and, as one staff member put it, "would tie us up for years."[69] The public was the only interest that got the short end of the stick. Thousands of chemicals approved by the EPA were fraudulently allowed onto the market and are still in use today.

Indeed, our entire approach to regulation is designed to benefit industry, not the public. For example, our regulatory agencies do not go out and challenge industry studies with independent scientific research. Rather, EPA

---

[65] E.G. Vallianatos with M. Jenkins, *Poison Spring*, ch. 7 (Bloomsbury Press 2014).

[66] Rebekah Wilce, "HOJO Transcript Illustrates EPA Collusion with Chemical Industry," *Independent Science News* (July 27, 2017); see also Transcript of Meeting Between EPA and Industry Representatives, October 3, 1978 at the Howard Johnson Hotel in Arlington Virginia, available at https://www.poisonpapers.org.

[67] Transcript of Meeting Between EPA and Industry Representatives, October 3, 1978 at the Howard Johnson Hotel in Arlington Virginia at 5, available at https://www.poisonpapers.org.

[68] Transcript of Meeting Between EPA and Industry Representatives, October 3, 1978 at the Howard Johnson Hotel in Arlington Virginia at 6-7, available at https://www.poisonpapers.org.

[69] Rebekah Wilce, "HOJO Transcript Illustrates EPA Collusion with Chemical Industry," *Independent Science News* (July 27, 2017); see also Transcript of Meeting Between EPA and Industry Representatives, October 3, 1978 at the Howard Johnson Hotel in Arlington Virginia, available at https://www.poisonpapers.org.

staff members are directed to summarily cut and paste corporate research studies into their chemical approvals without even looking at the underlying data or research methodologies provided by industry.[70]

The end result is that we are walking cesspools of toxic chemicals. The US Centers for Disease Control have conducted four major bio-monitoring studies since 1999 showing that Americans as a society are contaminated with hundreds of toxic chemicals such as BPA, phthalates, PFAS, pesticides, and heavy metals, among others.[71] These are chemicals found in our blood, urine, breast milk, etc. And not in just some of us, but in the vast majority— well over 90%—of us. Yet somehow their presence and impact on our bodies is not the center of our public health discussion. At the same time we have had an "alarming increase in serious illnesses in the US, along with a marked decrease in life expectancy" over the last couple of decades.[72]

The public knows very little about toxic chemicals used in commerce because industry protects details about them with confidential business information or "trade secret" claims. Companies can claim confidential business information about topics such as chemical identity, how much is in use, the physical location of the chemicals, and which products contain them.[73] This means that they do not have to disclose this information to the public and our regulatory agencies are bound to protect this information from us. The EPA has had neither the resources nor legal obligation to review the validity of industry's trade secret claims.[74] As a result, the vast majority have gone unchallenged and information about chemical toxicity falls through the cracks.[75]

---

[70] E.G. Vallianatos with M. Jenkins, *Poison Spring,* 6 (Bloomsbury Press 2014).

[71] US Centers for Disease Control, "Fourth National Report on Human Exposure to Environmental Chemicals" (2019) and previous reports dating back to US Centers for Disease Control, "National Report on Human Exposure to Environmental Chemicals (2001); see also European Union, "Endocrine Disruptors: from Scientific Evidence to Human Health Protection," a study commissioned by the PETI Committee of the European Parliament PE 608.866 (March 2019).

[72] Swanson, N.L., Leu, A., Abrahamson, J. & Wallet, B. "Genetically engineered crops, glyphosate and the deterioration of health in the United States of America," *Journal of Organic Systems* 9(2) 6–37 (January 2014).

[73] TSCA Chemical Data Reporting Requirements - Confidentiality Claims, 40 CFR §711.30; see also US EPA, "Freedom of Information Act (FOIA) - Limitations on Disclosure of Information under Pesticide Law," available at https://www.epa.gov/foia/limitations-disclosure-information-under-pesticide-law (last visited July 16, 2019) and Frank R. Lautenberg Chemical Safety for the 21st Century Act of 2016, Toxic Substances Control—Confidential Business Information, 15 USC §2613.

[74] See e.g., US EPA "Confidential Business Information Under TSCA—Frequent Questions about TSCA CBI," Response to Question 1, available at https://www.epa.gov/tsca-cbi/frequent-questions-about-tsca-cbi (last visited July 16, 2019).

[75] US EPA, "Confidential Business Information Under TSCA—Statistics for the TSCA CBI Review Program," available at https://www.epa.gov/tsca-cbi/statistics-tsca-cbi-review-

Unfortunately things are about to get worse. Until recently, companies had to prove that what they withheld as "confidential business information" was information sensitive to their competitive position. Going forward, thanks to a US Supreme Court decision,[76] it will be enough for chemical companies to designate any information as confidential to keep it from the public. The information no longer has to be sensitive to industry's competitive position to be kept private. It just has to be sensitive enough that industry does not want us to know about it. So any corporate documents indicating that a chemical is toxic can be withheld from the public going forward. Thank you, Supreme Court, for protecting industry from the public.

The reality is that whoever has most influence over the administrative, legislative and judicial branches of our government will control what the EPA will say and do. Industry—not the public—has much more influence in our political process. In fact, industry has a monumental advantage. It has exponentially more power—legally, politically, and financially—when it comes to influencing our politics, economy, and society. This power has grown to such an extent that the EPA—an agency tasked with protecting us from industry—is now an extension of industry.

The EPA is not the only regulatory game in town. We have state and other federal agencies who get involved in chemical regulation. These agencies are also captive to industry, however. And we have widespread industry use of toxic chemicals. This is true not only in the US but across the world, as measured by failed efforts to contain environmental pollution and protect human health.

---

program (last visited July 16, 2019). To see just what our confidential business information exemptions have had on the ability of industry to withhold information about toxic chemicals, take a look at the OECD database of the toxic PFAS class of chemicals (PFAS are the topic of Chapters 11-12). There are countless entries in the database where the US EPA could provide no information about the chemical volumes produced, imported, used, exported, or recycled in the US because the information has been withheld by the reporting company. 2018 OECD PFASs Spreadsheet, "Toward a New Comprehensive Global Database of Per- and Polyfluoroalkyl Substances (PFASs)" (OECD 2018).

[76] Food Marketing Institute v. Argus Leader Media, 588 US __ (June 24, 2019).

Chapter 6: The PR Industry

> All truth passes through three stages. First, it is ridiculed. Then, it is violently opposed. And finally, it is accepted as being self-evident.

> —Arthur Schopenhauer

While the EPA has been busy serving their corporate clients and industry has been busy making toxic chemicals, public relations (PR) companies have been busy shaping public opinion about our reality. PR companies are hired by industry to keep information about synthetic chemicals away from the public eye. PR firms create social media content and shape what ends up in the headlines. They dominate local and national news programs, radio talk shows, and all other modes of communication in the modern world. PR companies also enlist bloggers, internet trolls and people like you and me to create the content we see on Facebook, Twitter and social media.[77]

PR firms are effective because they invest their energies into understanding us and our psychology so that they can influence us easily. The art of influence has developed over many centuries and depends fundamentally upon understanding the audience. I have worked in some capacity helping

---

[77] John Stauber & Sheldon Rampton, *Toxic Sludge Is Good for You,* (Common Courage Press 1995); see also In Re: Roundup Products Liability Litigation, "Plaintiffs' Reply in Support of Motion to Strike Confidentiality of Heydens Deposition," Case No. 16-md-02741-VC, US District Court for the Northern District of California (April 20, 2017), referencing the deposition of David Heering and noting that "Monsanto even started the aptly-named "Let Nothing Go" program to leave nothing, not even Facebook comments, unanswered; through a series of third parties, it employs individuals who appear to have no connection to the industry, who in turn post positive comments on news articles and Facebook posts, defending Monsanto, its chemicals, and GMOs."

companies understand their audiences. It is one of the most valuable things a company can do to develop a successful product and secure market share. In an ideal world a company would use information about its audience in order to develop better products. However, this is not the reality of what usually occurs. Developing better products takes some effort and investment. It is much easier not to change anything and just tell people whatever they want to hear.

PR companies are hired to craft this messaging. Once they understand who we are, they enlist people who seem trustworthy to us in order to convince us of whatever they want us to believe. They choose people who talk the way we talk, dress the way we dress, can think the way we think, and appeal to our values. If we value individualism and property rights, they will enlist people who appeal to individualism and property rights. If we value inclusivity and diversity, they will enlist people who appeal to inclusivity and diversity. The messaging and delivery changes depending on the audience, but is ultimately tailored to get us on board with the idea that what industry does is good for us—or at least good enough.

Another important role of public relations is to suppress, obscure, rearrange, massage and distort any facts suggesting that there is a problem. The PR industry is known to resort to unethical conduct when it comes to this task.[78] Indeed, PR firms have engaged in industrial espionage, infiltrated civic and political groups, followed people, interfered with publications, harassed activists, and even incited mentally unstable individuals into violence so that the PR team could subsequently blame the violence on activist groups they do not like. The risk to PR companies should they get caught for such conduct are unfortunately relatively low. The financial reward for helping their clients sway public opinion, gain market share, and avoid legal liability are exponentially high.

There are some common tactics historically employed by the PR industry to sway public opinion. The first is to create doubt in the public eye that there is an actual danger—many times without outright denying the danger.[79]

---

[78] John Stauber & Sheldon Rampton, *Toxic Sludge Is Good for You* (Common Courage Press 1995); see also Interview of Carey Gillam and Gary Ruskin by Amy Goodman, "Documents Reveal Monsanto Surveilled Journalists, Activists & Even Musician Neil Young," *Democracy Now* (Aug. 9, 2019).

[79] John Stauber & Sheldon Rampton, *Toxic Sludge Is Good for You* (Common Courage Press 1995); European Environment Agency, "Late Lessons from Early Warnings: Science, Precaution, Innovation," ISSN 1725-9177 (2013); Gary Ruskin, "Seedy Business," US Right to Know (Jan. 2015), available at https://www.usrtk.org/seedybusiness.pdf; Cristin Kearns, "Sugar Industry and Coronary Heart Disease Research," *JAMA Internal Medicine* 176(11),1680-1685 (Nov 2016); Neela Banerjee, et al., "Exxon's Own Research Confirmed Fossil Fuels' Role in Global Warming Decades Ago" *Inside Climate News* (2015); Xavier Baur, et al., "Ethics, morality, and conflicting interests: how questionable profes-

This may include influencing what information we see when we search the internet, discrediting scientists and other experts, vilifying victims and objectors, and carefully crafting language to feed us half-truths we find to be palatable. Monsanto, for example, solicited prestigious academics from Harvard to disseminate its messaging about glyphosate (Chapter 10). The company also relied on bogus groups such as the "American Council on Science and Health" and the "Council for Biotechnology Information" to discredit scientists, journalists, and activists, and disseminate propaganda.[80]

It is common for industry to claim that we should not pay attention to animal studies because animals are different from people. What the public does not realize is that animal studies are universally accepted by the scientific community for establishing the anticipated health effects of toxic chemicals in our bodies.[81] Indeed, some of the most common scientific studies are performed on rats. Even chemical companies commonly utilize rat studies for the purpose of gaining regulatory approval. When a rat on other animal study comes out not in their favor, however, companies put out a PR spin claiming that animal studies are not valid.[82]

This technique is very effective because most of us do not want to believe that the things we buy or are exposed to in our everyday environment are harmful for us. Knowing that we are not safe would make us feel very uncomfortable. We do not want to feel uncomfortable, and so we are relieved by false assurances. We are thus satisfied by a short Google search and language from the first available authority that there are no definitive findings of harm to human beings, even if adverse animal studies exist. Indeed, our individual and collective response to evidence of any "awful truth" is to simply deny it.[83]

The EPA, FDA, and other respected organizations are frequently enlisted by industry to mislead the public.[84] For example, DuPont successfully got

---

sional integrity in some scientists supports global corporate influence in public health," *International Journal of Occupational Health* 21(2), 172–175 (March 2015); David Michaels, *Doubt is Their Product: How Industry's Assault on Science Threatens Your Health* (Oxford University Press 2008).

[80] Carey Gillam, *Whitewash*, 114-115, 121-122 (Island Press 2017).

[81] See e.g., "Revised Guidance Document 150 on Standardised Test Guidelines for Evaluating Chemicals for Endocrine Disruption," OECD, ISSN : 20777876 (Sept. 3, 2018).

[82] Ronald Melnick, et al. "War on Carcinogens: Industry Disputes Human Relevance of Chemicals Causing Cancer in Laboratory Animals Based on Unproven Hypotheses, Using Kidney Tumors as an Example," *International Journal of Occupational and Environmental Health* 19(4), 255-260 (Dec 2013).

[83] Derrick Jensen, *A Language Older Than Words*, 11, 65 (Context Books 2000).

[84] David Michaels, "Science for Sale," *Boston Review* (January 28, 2020); see also Carol Van Strum, *A Bitter Fog* (Jerico Hill Interactive 2014); E.G. Vallianatos, *Poison Spring* (Bloomsbury Press 2014); John Stauber and Sheldon Rampton, *Toxic Sludge is Good for You* (Common Courage Press 1995); Carey Gillam, *Whitewash* (Island Press 2017); Shiv Chopra, *Corrupt to the Core* (Kos Publishing 2009); "Corporate-Funded NGO Leads Industry Fight

the EPA to issue a public reassurance that Teflon products were safe during a media investigation into the safety of PFOA, a toxic chemical known to leech from Teflon frying pans during use and which had contaminated the Ohio River from DuPont's manufacturing facilities (Chapters 11–12).[85] A related PR tactic is to tell us that more data needs to be collected by the authorities before we can draw any conclusions or take any action. DuPont successfully used this tactic in litigation over Teflon.[86] It then made sure the authorities took no meaningful action. Such use of respected organizations is highly effective because while we may doubt what industry has to say, we are willing to believe our government, health organizations, and our professional peers.

Another common PR tactic used against us is to assure the public that the responsible authorities are taking care of things. This tactic is used when the scientific evidence and public concern start to overtake the corporate PR campaign in mainstream media. We, as the audience, may be shocked and dismayed to learn that a company is causing a major health and environmental problem with a toxic chemical or product. Naturally, we feel somewhat powerless to do anything about it. Therefore we are relieved to hear that the responsible company or regulatory agency is taking action. We generally let the event slide off of our radar and into media obscurity.[87]

However, in many cases very little is actually being done to address the public health and safety concern. Years may pass and nothing may change whatsoever. A few long and expensive studies may be published by regulatory agencies or industry groups along the way.[88] Consulting companies love working on such studies, as they pay the bills and lead to little controversy. Having worked in the consulting industry, I can tell you that the format of these studies is formulaic and designed to lull you into a sense of security. They take on a "nothing to see here" tone and mostly cut-and-paste from other meaningless reports. While these studies resolve very little, they get enough media attention to make us believe that someone cares and that progress is being made. The technique is very effective for any company who wants to get the public off of its back.

In case false statements of assurance do not work, the PR industry encourages us to disassociate from the victims of toxic chemical pollution

against Public Health Protection," *Sustainable Pulse* (June 5, 2019); Robert Bilott, *Exposure* (Simon & Schuster 2019).

[85] See Sharon Lerner, "The Teflon Toxin: Part 3," *The Intercept* (Aug. 20, 2015); the documentary by Stephanie Soechtig & Jeremy Siefert, *The Devil We Know* (2018), available at https://thedevilweknow.com/; and Robert Bilott, *Exposure* 220 (Simon & Schuster 2019).

[86] Robert Bilott, *Exposure*, 284–285 (Simon & Schuster 2019).

[87] Robert Bilott, *Exposure* ch. 29 (Simon & Schuster 2019).

[88] Robert Bilott, *Exposure*, 284–285 (Simon & Schuster 2019).

and our own health problems in order to maintain our sense of security. We are consistently fed the story that people who have fallen ill with a life-threatening illness—cancer, heart disease, diabetes—must have done so out of their own accord. Either it was in their genes, individual lifestyle choices or stress that did them in. Victims with more subtle and unexplainable chronic problems—such as chronic fatigue and chemical sensitivity—are characterized as hysterical or emotionally unstable.

For this reason, even doctors are frequently unsupportive to the victims, believing their health problems must be caused by lifestyle or emotional problems. This is despite the decades of scientific research showing the connection between various chemicals and human health outcomes (Chapters 8-23). This research never makes it into our educational system, however—be it middle school or medical school. Instead, news stories, health websites, wellness gurus, the pharmaceutical industry and an entire army of PR staff are deployed to convince us that the status of our health and wellbeing in a highly polluted world is exclusively our own fault.

While we are getting these official messages from formal media channels, the PR industry is also busy attacking the scientists, whistleblowers, health and environmental advocacy groups, and other dissenters. Their focus is on discrediting the individuals or groups with any means necessary. It is common for chemical companies to attack people who conduct studies unfavorable to them for the purpose of discrediting them and pressuring them to cease their research. They may even threaten a person's safety or the safety of their families. The PR machine has no reservations about trying to ruin a person's career and reputation when profits are at stake.[89]

For example, recent evidence shows that Monsanto kept files on over 200 journalists and lawmakers in France for the purpose of influencing the country's position on pesticides.[90] Upon further investigation, it was found that Monsanto kept watch lists on journalists and lawmakers across Europe—including those in individual European countries and the European Union in general.[91] Months later it was discovered that a similar PR war was being waged by Monsanto in the United States—the company dedicated an entire

---

[89] Jeffrey M. Smith, "GMO Researchers Attacked, Evidence Denied, and a Population at Risk," Global Research Centre for Research on Globalization," *Global Research* (Sept. 19, 2012).

[90] Ludwig Burger, "Monsanto May Have Kept Private Data on Europeans to Influence Pesticide Views," *Insurance Journal* (May 15, 2019).

[91] "Monsanto Kept 'Watch Lists' in Seven EU Countries – Bayer," *Sustainable Pulse* (May 21, 2019).

"Fusion Center" utilizing a team of former intelligence agents to keeping dibs on journalists and others who publish unfavorable material.[92]

As a group, I have noticed that women become a special target of industry PR campaigns. Women tend to be the first to object when corporate or government actions harm human health or the environment. Women are primary caregivers for others, and are therefore more deeply affected when the health of those around them—their children, parents, spouses—are suffering. The health of their family is a high priority and they are motivated to understand the connection between that health, society, and economy. When women speak up about the health dangers of toxic chemicals, they are dismissed as irrational and hysterical, even when the scientific evidence is clearly on their side.

It is not possible for us to dismiss the dissenting voices of others unless we consent to it. We as individuals are told by the PR industry to not listen. Leveraging our biases, fears, and stereotypes, they encourage us to doubt. As such, rural communities are perceived to be filled with dumb rednecks. Nature-loving liberals are viewed as crazy tree-huggers. Wealthy suburbanites are characterized as NIMBYs. The poor, minorities, developing world or, more broadly, people different from us are judged as having themselves to blame.

We are vulnerable to these various PR techniques—dismissing the victims, discrediting the science, doubting the dangers, and believing that all is well or being taken care of. This is because knowing the truth would be very painful and overwhelming and question our sense of reality. It would require us to face the fact that the pillars of society are not keeping us safe and secure. That what we are taught in college, on TV, or at work is not an accurate representation reality. That no political party, democratic or republican—or any other entity with authority—has looked out for our best interests. And that we have been duped by a PR campaign that has convinced us that our success, productivity, sexual attractiveness, and happiness depend upon the consumption of products that are in fact poisoning us and destroying our environment.

---

[92] Interview of Carey Gillam and Gary Ruskin by Amy Goodman, "Documents Reveal Monsanto Surveilled Journalists, Activists & Even Musician Neil Young," *Democracy Now* (Aug. 9, 2019).

Chapter 7: Monopolizing Our World

The chemical industry has grown its sales to more than $5 trillion globally.[93] It dominates agribusiness, chemical production, new pharmaceutical drug development, and biotechnology in general. One way in which the industry has gained such power has been to engage in corporate mergers. For example, the pesticide sector, with $35 billion in annual sales, has consolidated substantially over time and is now dominated by a small handful of manufacturers.[94] Its concentration of market power has grown considerably with the consolidations of Monsanto-Bayer, Syngenta-ChemChina and Dow-DuPont.[95]

Industries also use nefarious means of gaining market power. Corrupting foreign governments is one such strategy.[96] As a result of research done by the International Consortium of Investigative Journalists, we have a mountain of evidence of international companies paying off government officials to pass laws that give them unfettered market access.[97] Corporations are

---

[93] "Global Chemicals Outlook II," United Nations Environment Programme, ISBN No: 978-92-807-3745-5 (2019) (utilizing 2017 data).

[94] "Global Chemicals Outlook II," United Nations Environment Programme, ISBN No: 978-92-807-3745-5 (2019).

[95] "Global Chemicals Outlook II," United Nations Environment Programme, ISBN No: 978-92-807-3745-5 (2019).; DuPont has since been restructured. See "DowDuPont Completes Spin-off of Dow, Inc.," DuPont (April 1, 2019), available at https://www.dupont.com/news/dowdupont-completes-spin-off-of-dowinc.html.

[96] See generally Vandana Shiva, *Oneness vs. the 1%* (Spinifex Press 2018).

[97] See e.g., Will Fitzgibbon, "The Panama Papers: Secret Offshore Deals Deprive Africa Of Billions In Natural Resource Dollars," *International Consortium of Investigative Journalists* (July 25, 2016).

known to present cash under the table to obtain government contracts and overcome local resistance to their presence.[98]

The manipulation of international trade is also a long-standing contributor to the growth of corporate market power. International institutions—such as the IMF, World Bank, WTO, and divisions of the UN—become the target for corporate influence. These institutions shape international investment and policy. Successful lobbying of them can secure access to market opportunities and power for industry where it would not otherwise exist.[99] As a result of corporate lobbying most international trade agreements—including GATT, NAFTA, TRIPS, and others—are designed to protect the interests of a few major corporations while suppressing the ability of national governments and local peoples to secure their health, environmental, and economic interests.[100]

Big Ag is the poster child of corporate influence on international trade. It has pushed out small organic farmers across the world and made global food production largely dependent upon pesticides, proprietary seeds, and industrial-scale farming technologies. The pesticide industry effectively uses US free trade agreements to secure monopolies in developing nations while destroying existing local economic activity. It is common for US free trade agreements to contain a restrictive clause effectively giving a 10-year monopoly in developing nations to one of the major international agrochemical companies.[101] The effect of such agreements is not free trade, but rather growing market power for the global agrochemical oligopoly.

In Colombia, the lucky beneficiary of such "free trade" agreements has been Monsanto. Monsanto was chosen under a US/Colombia agreement to aerial spray the countryside of Colombia with glyphosate to purportedly

[98] See e.g., Sasha Chavkin, "Leak Exposes Millions Of Dollars In New Payments In Odebrecht Cash-For-Contracts Scandal," International Consortium of Investigative Journalists Bribery Division (June 25, 2019); and Fergus Shiel, & Sasha Chavkin, "Bribery Division: What Is Odebrecht? Who Is Involved?" International Consortium of Investigative Journalists Bribery Division (June 25, 2019).

[99] See e.g., AccountAbility, "Towards Responsible Lobbying," United Nations Global Compact (2005); and "The COP19 Guide to Corporate Lobbying," Corporate Europe Observatory (2013).

[100] Interview with Vandana Shiva in "20 Years After The Battle of Seattle: Vandana Shiva & Lori Wallach on Historic 1999 WTO Protests," *Democracy Now* (Nov. 27, 2019); see also "More Information on Investor-State Dispute Settlement," Public Citizen's Global Trade Watch available at https://www.citizen.org/article/more-information-on-investor-state-dispute-settlement (last visited Dec. 2, 2019); see also Trans-Pacific Partnership, Article 18.47 (the US is not a current signatory).

[101] See e.g., United States-Columbia Trade Promotion Agreement, art. 16.10 (May 15, 2012), United States-Mexico-Canada Agreement, art. 20:45 (Nov 30, 2018); Dominican Republic-Central America Free Trade Agreement, art. 15:10 (Aug 5, 2004) (this agreement was subsequently applied to Costa Rica, El Salvador, Guatemala, Honduras and Nicaragua as well).

reduce cocaine production.[102] Over a twenty-year time period, the spraying destroyed legitimate crops and farmland owned by small farmers. It also led to a huge rise in birth defects and human health problems. Monsanto's operations in Colombia did little to actually curtail the drug industry.[103] They did help Monsanto establish a foothold in Colombia's agricultural market, however.

The Colombian government helped Monsanto even further, standing good on its commitment to big agriculture under its free trade agreement with the US.[104] The country passed a law making it illegal for farmers to save their seeds, thereby forcing farmers to purchase seeds from Monsanto. Police harassed entire communities, destroyed local grain supplies, and prosecuted unsuspecting farmers who continued to keep their seeds. Civil revolt ensued, and fortunately three years later the law was retracted.[105] The damage had been done, however; Colombia is now infiltrated by the global pesticide industry and its local farming techniques and economies have been destroyed. Other farmers around the world have had similar experiences with government-sanctioned colonization by the global chemical companies.[106]

The agrochemical industry is increasingly also hiding behind non-profits to further its economic interests around the world.[107] Engaging non-profits is especially problematic because it creates the false impression that corporations are do-gooders. For example, agrochemical companies have been looking for a way to unleash a new product—mosquitoes that will sterilize other mosquitos (aka exterminator technologies). They argue that this will help control the spread of disease. Much of the international community disagrees, as the ecological impacts of setting loose sterilizing mosquitoes

---

[102] Compare Sibylla Brodzinsky, "Last flight looms for US-funded air war on drugs as Colombia counts health cost," *The Guardian* (May 6, 2015) and Elena Sharoykina, "Monsanto: The Pentagon's Soldier in Colombia" *Sustainable Pulse* (Nov 15, 2016) with United States-Columbia Trade Promotion Agreement, art. 16.10 (May 15, 2012).

[103] Adam Isacson, "Restarting Aerial Fumigation of Drug Crops in Columbia is a Mistake," WOLA Analysis (March 7, 2019).

[104] The US/Columbia Free Trade Agreement utilizes patent protection over plants/seeds and market exclusivity clauses to secure monopoly power to large agricultural chemical companies. United States-Columbia Trade Promotion Agreement, art. 16.9(2) & 16.10(1) (May 15, 2012).

[105] "Colombia farmers' uprising puts the spotlight on seeds," GRAIN (Sept 4, 2013), available at https://www.grain.org/article/entries/4779-colombia-farmers-uprising-puts-the-spotlight-on-seeds ; and Clementina Films, "9.70 documentary," directed by Victoria Solano, published on YouTube on October 1, 2013.

[106] See Glyn Moody, "Guatemala Resists 'Monsanto Law' Required As Part Of Trade Agreement With US," *TechDirt* (Sept. 3, 2014); and Glyn Moody, "Africa's Ancient Plant Diversity And Seed Independence Under Threat, Supposedly In The Name Of Progress," *TechDirt* (May 1, 2013).

[107] Vandana Shiva, *Oneness v. The 1%* (Spinifex Press 2018).

into the environment are unknown and could be disastrous.[108] Nevertheless, with the help of a non-profit funded by the Bill & Melinda Gates Foundation, big agribusiness moved forward with testing of these exterminator technologies in Burkina Faso, a small country in Africa. This testing occurred without permission from the local communities.[109] The chair of the Alliance for Food Sovereignty in Africa released a statement objecting to this corporate practice, reflecting the public's concern: "In Africa we are all potentially affected and we do not want to be lab rats for this exterminator technology."[110]

Exterminator technologies are one entry on a growing list of dubious technologies focused around the genetic modification of organisms (GMO). GMO technologies enable industry to profit from the manipulation of life and ultimately give corporations the capacity to monopolize entire economic sectors. This capacity is inadvertently created by our intellectual property laws. Intellectual property laws are intended to encourage the development of new technologies. They give protection to those who invest in technology development and prevent opportunists from copying innovations during a grace period. However, intellectual property laws are being abused to secure monopoly power for the agrochemical industry.

For example, GMO seeds are designed to produce crops that are either bred to be herbicide resistant or to contain pesticidal toxins.[111] The most common herbicide resistant plants are "RoundUp ready," or plants that are resistant to glyphosate. RoundUp-ready crops absorb the glyphosate herbicide sprayed onto the plant but do not die. We, in turn, absorb glyphosate herbicide when we eat these crops (see Chapter 10). GMO seeds were an ingenious way for industry to get around an international moratorium on "terminator seeds," which were sterile seeds that pesticide companies

---

[108] "Genetic extinction technology challenged at UN Convention on Biological Diversity," (Nov. 18, 2018), Friends of the Earth International Press Release, available at https://www.foei.org/news/genetic-extinction-technology-challenged-at-un-convention-on-biological-diversity; and "A Human Rights Analysis of Gene Drives," FIAN International Secretariat (Nov. 2018), available at bch.cbd.int/database/attachment/?id=18977.

[109] "Expert Groups Slam Release of GM Mosquitoes in Burkina Faso," *Sustainable Pulse* (July 6, 2019).

[110] Quote from Mariann Bassey-Orovwuje, Chair of the Alliance for Food Sovereignty in Africa, published in Joel Achenbach, "'Gene drive' research to fight diseases can proceed cautiously, U.N. group decides," *The Washington Post* (Nov. 30, 2018). As a result of lobbying during the United Nations Convention on Biological Diversity held in Egypt in November 2018, exterminator technologies have been approved. UN Convention on Biological Diverstiy, "Decision Adopted by the Parties to the Cartagena Protocol on Biosafety," CBD/CP/MOP/DEC/9/13 (Nov. 30, 2018); see also Kerry Grens, "No Ban on Gene Drives: UN Convention - United Nations members agree to some restrictions on the technology, but not a total suspension of its use." *The Scientist* (Nov. 30, 2018).

[111] US Department of Agriculture, "Recent Trends in GE Adoption," (updated July 16, 2018), available at https://www.ers.usda.gov/data-products/adoption-of-genetically-engineered-crops-in-the-us/recent-trends-in-ge-adoption.aspx.

---

wanted to sell back to farmers year after year.[112] Terminator seeds would have given 100% market power to pesticide companies. Farmers would have had no way to plant their own crops. As such, terminator seeds would have given Big Ag complete power over the global food supply.

Terminator seeds were largely rejected by the international community because they were such a threat to food security. GMO seeds are completely legal, however. And unfortunately they leave the farmer reliant upon big agriculture. Once a farmer accepts GMO seeds from a pesticide manufacturer, the farmer becomes reliant upon the pesticide company, including all of the products and services designed specifically to be used with GMO seeds. In addition, the farmer must continue to buy the seeds or pay royalties because the seeds are patented and protected by intellectual property laws.[113] The farmer can never stop using the seeds because the entire crop will inevitably become germinated by the GMO strain during the next pollination season. The farmer unwittingly becomes a captive customer for decades.

The likelihood that neighboring farms will also become germinated—and therefore contaminated—is very high.[114] Should a farmer's crops become contaminated by GMO crops, the cost of remedying the situation will fall on the farmer.[115] The farmer is also at risk of being sued by pesticide manufacturers for possessing GMO crops and failing to pay royalties; such litigation is likely to be so costly—even when it is baseless—that it can bankrupt a farm.[116] All of this creates a difficult business proposition for small and organic farmers around the globe. The high probability and costs associated with GMO and pesticide contamination puts them at an economic risk of becoming an endangered species, even if consumers prefer to buy local or organic produce.

As of 2018, over 90% of corn, cotton and soy beans produced in the United States were GMO crops.[117] If not sold to us directly, GMO corn and

---

[112] See Convention on Biological Diversity, "Genetic Use Restriction Technologies (GURTs)," available at https://www.cbd.int/agro/gurts.shtml (last visited June 27, 2019). Terminator seeds are formally known at Genetic Use Restriction Technologies.

[113] Bowman v. Monsanto, 569 U.S. 278 (2013).

[114] Danny Hakim, "Monsanto's Weed Killer, Dicamba, Divides Farmers," *The New York Times* (Sept. 21, 2017); Eric Lipton, "Crops in 25 States Damaged by Unintended Drift of Weed Killer," *The New York Times* (Nov. 1, 2017).

[115] Danny Hakim, "Monsanto's Weed Killer, Dicamba, Divides Farmers," *The New York Times* (Sept. 21, 2017); Eric Lipton, "Crops in 25 States Damaged by Unintended Drift of Weed Killer," *The New York Times* (Nov. 1, 2017).

[116] Chris Womack, "Monsanto Attacks!" *Texas Observer* (Nov. 9, 2001); Schmeiser v. Monsanto, 1 S.C.R. 902 (2004); "Monsanto vs. U.S. Farmers" (2005), Center for Food Safety, available at https://www.centerforfoodsafety.org/files/cfsmonsantovsfarmerreport11305.pdf

[117] US Department of Agriculture, "Adoption of Genetically Engineered Crops in the U.S.: Genetically engineered varieties of corn, upland cotton, and soybeans, by State and for

soy are fed to the animals we eat. We in turn eat these animals. Labeling of GMO crops is not required in the US as a result of heavy lobbying by pesticide companies. These companies know that many consumers would not buy GMO if it was disclosed on the label. They have thus spent large sums of money and countless hours defeating local citizen initiatives that would require the labeling of GMO crops.[118] As a result, organic farmers now bear the cost of paying for expensive GMO-free and organic certifications.[119] This has the effect of further marginalizing organic farmers in the marketplace.

GMO technologies are problematic enough, but things are about to get even more problematic. The agrochemical industry is working to introduce pesticides with special technologies that penetrate the cells of the body of insects and mutate their genetic material.[120] The purpose of these mutations is to disrupt the proper functioning of the cells so that the insect will die or fail to reproduce. Spraying fields with such pesticides can put all insects visiting a given area at risk, including essential pollinators such as honeybees.

Regional studies reveal that there has already been more than a 70% decline in common types of insects in less than three decades.[121] Gene mutating pesticides may accelerate the species loss and could put human health at further risk. Should these gene-mutating technologies get out of hand and adversely affect agriculture, the pesticide industry will likely turn around and sell GMO solutions at a steep price to address the problem. If these gene mutating technologies adversely affect humans, the pharmaceutical branches of these same companies will make a steep profit from drugs to restore our health.

Gene mutating pesticides are a product of an alliance between the agrochemical and biotechnology industries. These types of cross-sector alliances are becoming increasingly more common. Big Ag is also working with Big Tech.[122] Technology is increasingly being used to gather data on crops, livestock, seeds, and even weather conditions. While these services may at first blush seem helpful to agriculture, they serve as an additional way to consoli-

---

the United States, 2000-19," available at https://www.ers.usda.gov/data-products/adoption-of-genetically-engineered-crops-in-the-us.aspx (September 18, 2019).

[118] Libby Foley, "Corporate Spending to Fight GMO Labeling Skyrockets," *Environmental Working Group* (April 23, 2015); and Rob Coleman, "Food Lobby Spends $101 Million in 2015 to Avert GMO Labeling," *Environmental Working Group* (February 25, 2016).

[119] See e.g., USDA "Becoming a Certified Operation - FAQs: How Much Does Organic Certification Cost?," available at https://www.ams.usda.gov/services/organic-certification/becoming-certified (last visited June 27, 2019).

[120] "Gene-Silencing Pesticides: Risks and Concerns," Friends of the Earth (Oct. 2020)

[121] Francisco Sanczez Bayo & Kris A.G. Wyckhuys, "Worldwide decline of the entomofauna: A review of its drivers," *Biological Conservation* 232, 8-27 (April 2019).

[122] Vandana Shiva, *Oneness v. The 1%* (Spinifex Press 2018).

date farming operations for Big Ag at the expense of small farmers. Activists such as Vandana Shiva are raising the flag on the tracking of farming operations, legitimately concerned about the harvesting of data as a form of corporate surveillance and therefore a monopolization tool:

> The intelligence of nature and people is being replaced by 'intelligence' as surveillance—whether surveillance by Monsanto over farmers to prevent them from saving seeds, or surveillance by Facebook and Google....[123]

The European Union agrees with this sentiment—at least when it comes to surveillance of the average citizen; it is investigating Big Tech for utilizing its customer data to monopolize the marketplace.[124] The use of surveillance by Big Ag has yet to be formally evaluated for corporate abuse.

Big Ag, like many other industries, utilizes carefully crafted language to garner public support for its efforts. The widespread deployment of pesticides and GMO crops around the globe has been characterized by pesticide companies as "sustainable agriculture" that promotes "global food security." While both sustainable agriculture and global food security are admirable goals, they are unlikely to be achieved though the elimination of small farms, poisoning of the world, and monopolization of economies around the globe.

---

[123] Vandana Shiva, *Oneness v. The 1%*, 24 (Spinifex Press 2018).

[124] Elizabeth Schulze, "If you want to know what a US tech crackdown may look like, check out what Europe did," *CNBC.com* (June 7, 2019); Jason Del Rey, "Why Congress's antitrust investigation should make Big Tech nervous," *Vox* (Feb. 6, 2020).

CHAPTER 8: DESIGNED TO KILL LIFE

Pesticides are one of the deadliest and most ubiquitous series of chemicals that circulate in the air, water, soil and food systems. Insecticides, herbicides, fungicides, and germicides are *biocides*—that is, they are designed to kill life. Their fundamental purpose is to destroy what nature has built—weeds, insects, and other creatures that are a part of nature's biodiversity.[125] And pesticides harm more than the intended targets. They create a whole host of casualties among plants, animals, people, and even unborn human beings. When they do not kill, they maim and encourage other diseases to thrive. They soak into the soil and our food. They drift in the air to our communities. And they contaminate our water. Their footprints are felt throughout our ecosystem, and cause our soil quality to degrade, bird and fish populations to plummet, beneficial pollinators to disappear, and drug resistant diseases to spread.

The realities of pesticides were brought to light by Rachel Carson.[126] The devastating impacts of these chemicals on our ecosystem have only grown over time, undermining the survival of up to one million species (25% of life on Earth) in the next few decades.[127] No one is untouched by their path of destruction; but most people do not realize their impact.

---

[125] Pesticides is an umbrella term that includes insecticides, herbicides, fungicides, and biocides.

[126] Rachel Carson, *Silent Spring* (The Riverside Press 1962).

[127] S. Díaz, et al., "Summary for policymakers of the global assessment report on biodiversity and ecosystem services of the Intergovernmental Science-Policy Platform on Biodiversity and Ecosystem Services," United Nations, Intergovernmental Science-Policy Platform on Biodiversity and Ecosystem Services (IPBES) (May 6, 2019) available at https://ipbes.net/global-assessment.

Thanks to the documentation provided in *Silent Spring*, we know that before the mass introduction of pesticides in the 1950s things looked very different in the US. The rivers were teeming with fish. Wildfowl and game roamed the land. Wildflowers bloomed along roadways and in fields. Birds flocked in trees like Christmas ornaments, and their song filled the air in the springtime. This all changed. In a systemic effort orchestrated by pesticide manufacturers and facilitated by government agencies, the US began a pesticide and herbicide spraying program around the country. The majority of our wildlife was decimated between the 1950s and 1970s. Spring grew silent. The song of birds and sounds of wildlife no longer filled the air with the changing of the seasons.

There was always some stated reason to spray. To reduce vegetation and increase visibility on roadways. To eliminate mosquitos. To control plant diseases. The list went on. As a result, birds fell from the sky *en masse* and lay dead in fields and suburban backyards. Entire populations of fish floated dead in the rivers. Insects and species of small mammals disappeared from many landscapes. Wildlife was born with deformities or was found still-born by those who ventured into the forest.

People were shocked by this, outraged, and often times left traumatized and hysterical. They wrote to their local newspapers. They went to their political representatives. But not much changed. Rachel Carson wrote *Silent Spring* as a result of these mass exterminations of life. Her work caused environmental laws to be passed in Washington DC and started the environmental movement across America. With the passage of these laws, Americans thought our government had taken care of the problem. But in reality, the onslaught has continued from decade to decade.

Pesticides used in the US today cause chronic health effects at low-levels of exposure, creating long-term problems such as fatigue, chronic pain, muscle weakness, insomnia, irritability & neurological changes, headaches, Parkinson's disease, and cancer, among other forms of disease.[128] In children, low-level exposures to pesticides are implicated in the development of autism and ADHD as well as cancer.[129] Some of these chemicals are strong

---

[128] James R Roberts & J Routt Reigart, *Recognition and Management of Pesticide Poisonings: Sixth Edition*, "Chapter 21: Chronic Effects" (2013), available from the USEPA Office of Pesticide Programs; see also Freya Kamel & Jane A. Hoppin, "Association of Pesticide Exposure with Neurologic Dysfunction and Disease," Environmental Health Perspectives, 112(9), 950–958 (June 2004).

[129] James R. Roberts & J. Routt Reigart, *Recognition and Management of Pesticide Poisonings: Sixth Edition*, "Chapter 21: Chronic Effects" (2013) available from the USEPA Office of Pesticide Programs; see also Freya Kamel & Jane A. Hoppin, "Association of Pesticide Exposure with Neurologic Dysfunction and Disease," *Environmental Health Perspectives*, 112(9), 950–958 (June 2004).

hormone disruptors (see Chapters 15–17); they are known to cause reduced semen count and genital birth defects in males, early puberty and fertility issues in females, changes in sexual behavior in both genders, and thyroid problems, among others.[130]

Pesticides are problematic not only when we ingest them from the fruits, grains and vegetables we eat. They also leach into our waterways and soil where they are absorbed by the plants and animals exposed to them (including unsuspecting human beings). We not only eat produce tainted with pesticides, but also fish, poultry, livestock and other species that have been contaminated with these chemicals. We drink water and swim in rivers, lakes, and bays contaminated with pesticides. And sometimes that contamination starts in agricultural fields and forests hundreds of miles away from where we live—way outside of our conceptual radar.

The EPA sets "tolerance levels" for how much pesticides should safely be in our food and environment. This gives many people the false impression that the dose makes the poison and even if there is a small dose, it is not poisonous if the EPA has approved the dose. This might be true in theory for some chemicals; however the EPA's tolerance levels do not reflect a safe dose. Rather they are "adjusted" to protect industry.[131]

If pesticide contamination is currently at a given level, the EPA sets tolerance levels above that. If the legal tolerance level dropped below the actual level of contamination in our food, soil, water, air, blood, and urine, then companies would be exposed to legal liability. And companies do not want that. So they make sure the EPA always stays ahead of the curve on higher tolerance levels to keep industry safe.[132] As a result of this approach designed to protect industry, acceptable glyphosate residue levels in food vary extensively—from less than .1 ppm (part per million) to over 300 ppm—depending on the type of food (e.g., almonds, meat, corn, wheat, etc.) and its level of contamination.[133] It is only when a company has truly fallen from grace in the public's eye that the EPA feels politically motivated to start lowering tolerance levels below the status quo.

[130] James R Roberts & J Routt Reigart, *Recognition and Management of Pesticide Poisonings: Sixth Edition,* "Chapter 21: Chronic Effects" (2013) available from the USEPA Office of Pesticide Programs; see also Freya Kamel & Jane A. Hoppin, "Association of Pesticide Exposure with Neurologic Dysfunction and Disease," *Environmental Health Perspectives,* 112(9), 950–958 (June 2004).

[131] E.G. Vallianatos with M. Jenkins, *Poison Spring,* 16–21, 83 (Bloomsbury Press 2014).

[132] E.G. Vallianatos with M. Jenkins, *Poison Spring,* 16–21, 83 (Bloomsbury Press 2014).

[133] USDA, "Questions and Answers on Glyphosate," (updated 10/31/2018), available at https://www.fda.gov/food/pesticides/questions-and-answers-glyphosate; Glyphosate; tolerance for residues, 40 CFR 180.364; see also Marek Cuhra, "Review of GMO safety assessment studies: glyphosate residues in Roundup Ready crops is an ignored issue," *Environmental Sciences Europe* 27, 20 (Sept. 2015) (original figures presented in mg/kg).

The sheer volume of pesticides used is also a tip-off for the extent of the problem. The US produces over one billion pounds of "active" ingredients used in pesticides annually.[134] This is enough to spray 2.5 pounds per acre of cropland on US soil.[135] And these figures reflect only the "active" pesticide ingredients listed for regulatory purposes. Active ingredients are mixed with many "inert" ingredients that frequently make these chemical cocktails even more poisonous.[136] Former EPA whistleblower EG Vallianatos explains that the term "inert" is a legal designation and misnomer. "Inert" ingredients are those that the pesticide companies do not have to register with regulatory agencies and can make up 99% of any given formula. They may include ingredients that will make the active ingredient more powerful, penetrate plant and animal cells more persistently, or simply make the entire formula more toxic. Indeed, inert ingredients may include cancer-causing petroleum distillates, neurotoxins, industrial solvents, and even substances otherwise banned from commercial use.[137]

You might wonder whether the majority of the pesticides produced here are exported, thus explaining the sheer volume of production—but the reality is that the US produces less than 18% of the world's pesticides. Other regions of the world—Europe, Asia, South America, etc., are making their own. Globally, approximately 6.26 billion pounds of pesticides are produced annually.[138] That is enough to apply over 1.5 pounds of active pesticide ingredient for each acre being farmed.[139]

The use of pesticides took off largely as a result of World War II, when many of the chemicals were developed for use in chemical warfare and wartime disease control. Most people are familiar with DDT, an insecticide pushed heavily onto the global market after its use in World War II. DDT is an organochlorine, a highly persistent, bio-accumulative, and toxic class of insecticides. It is stored in our fat tissue and lasts generations. Over 2 million metric tons of DDT was produced globally, primarily between the 1950s and

---

[134] "Global Chemicals Outlook II" 46, ISBN No: 978-92-807-3745-5, United Nations Environment Programme (2019).

[135] Daniel P. Bigelow and Allison Borchers, "Major Uses of Land in the United States, 2012," United States Department of Agriculture (2017).

[136] Robin Mesnage, et al., "Major Pesticides Are More Toxic to Human Cells Than Their Declared Active Principles," *Biomed Research International* 2014, 179691 (Feb. 2014).

[137] E.G. Vallianatos with M. Jenkins, *Poison Spring*, 23-29 (Bloomsbury Press 2014); see also Robin Mesnage, et al., "Major Pesticides Are More Toxic to Human Cells Than Their Declared Active Principles," *Biomed Research International* 2014, 179691 (Feb. 2014).

[138] "Global Chemicals Outlook II" 46, ISBN No: 978-92-807-3745-5, United Nations Environment Programme (2019).

[139] "The State of the World's Land and Water Resources for Food and Agriculture," Food and Agriculture Organization of the United Nations (2013).

1970s.[140] This was one of the pesticides sprayed across the country that motivated Rachel Carson to write *Silent Spring*.

The chemical was also used globally in hospitals to sterilize medical equipment.[141] The use of DDT in hospitals overlapped with a series of health outbreaks that affected hospital staff and nurses around the world. The most famous such outbreak occurred in the Royal Free Hospital in London. Staff who fell ill initially experienced fatigue, drowsiness, muscle pain, severe headache, vomiting, and low grade fever. After a few days, the muscle pain became more severe, turned into muscle weakness and in some cases paralysis. Many staff members were still experiencing chronic fatigue one year later and developed neurological problems. A few had symptoms of partial paralysis years onward.[142] No pathogen was ever found.[143] It is highly probable that DDT was the culprit; DDT exposure is associated with the same symptoms experienced by the Royal Free Hospital staff, including fatigue, headache, muscle weakness, neurological problems, and paralysis.[144]

Even though DDT has largely been banned in the US as of the late 1970s, other similar organochlorines remain in active use.[145] Moreover, DDT and other organochlorines continue to be widely found in blood samples of Americans today.[146] Exposure to organochlorine insecticides can lead to a variety of symptoms—ranging from short-term muscle weakness to polio-like symptoms that lead to permanent paralysis.[147] More benign symptoms include headache, nausea, lack of coordination, dizziness, and mental

---

[140] "Global Chemicals Outlook II" 47, ISBN No: 978-92-807-3745-5, United Nations Environment Programme (2019).

[141] US Department of Interior, "DDT Health and Safety Update," *Conserve O Gram Newsletter* 2(14) (Dec 2000), available at https://www.nps.gov/museum/publications/conserveogram/02-14.pdf.

[142] "Royal Free Disease, Sixty Years On," Summary of talk given at the Royal Free Association meeting, 20 November 2014 by Rosemary Underhill and Rosemarie Baillod, available at http://iacfsme.org/PDFS/Newsletter-Attachments/Royal-Free-disease-abstract-%28attachment-2%29.aspx.

[143] "Royal Free Disease, Sixty Years On," Summary of talk given at the Royal Free Association meeting, 20 November 2014 by Rosemary Underhill and Rosemarie Baillod, available at http://iacfsme.org/PDFS/Newsletter-Attachments/Royal-Free-disease-abstract-%28attachment-2%29.aspx.

[144] US Department of Interior, "DDT Health and Safety Update," *Conserve O Gram Newsletter* 2(14) (Dec 2000), available at https://www.nps.gov/museum/publications/conserveogram/02-14.pdf.

[145] James R Roberts & J Routt Reigart, *Recognition and Management of Pesticide Poisonings: Sixth Edition*, "Chapter 7: Organochlorine Pesticides" (2013), available from the USEPA Office of Pesticide Programs.

[146] "Fourth National Report on Human Exposure to Environmental Chemicals" US Centers for Disease Control (2019).

[147] James R Roberts & J Routt Reigart, *Recognition and Management of Pesticide Poisonings: Sixth Edition*, "Chapter 5: Organophosphate Pesticides" (2013), available from the USEPA Office of Pesticide Programs.

confusion.[148] DDT specifically is also known to severely weaken the shells of birds—to such an extent that many would crack under the pressure of roosting birds before the chicks could hatch.[149] It is believed that this occurs due to their hormone disrupting effects (see Chapters 15–17). Next time you are at the grocery store, buy the most organic, free-range eggs you can find, as well as the most conventional and cheapest eggs you can find. Go home and crack the eggs—chances are very high that the conventional eggs will crack much more easily than the organic eggs. The difference in the shells suggests that conventional chickens continue to be exposed to organochlorines or other hormone disruptors.

Organophosphates, another class of insecticides, took over the market after organochlorines fell out of popularity. Between 2000 and 2012, Americans went through 456 million pounds of organophosphates.[150] They continue to be used as pesticides in the marketplace today.[151] Organophosphates—like organochlorines—damage the central nervous system. This damage occurs regardless of whether the target species is a human, another mammal, a bird, insect or amphibian. Like with all chemical exposures, symptoms of organochlorine poisoning may not appear right away. They may develop over weeks or months, and may even develop as a disease years later, as these chemicals can cause subtle but permanent damage in the body.[152]

Organophosphates interfere with an essential neurotransmitter, causing a wide range of health problems in the long-run. Symptoms of exposure can express themselves as headache, nausea, sensory disturbances, lack of physical coordination, muscle weakness, muscle twitching, abdominal cramps, allergy-like symptoms (e.g., watery eyes, runny nose), neuropathy, anxiety, confusion, memory loss, blurred vision, excessive salivation, impaired breathing and/or in the worst case, paralysis. Children exposed

---

[148] James R Roberts & J Routt Reigart, *Recognition and Management of Pesticide Poisonings: Sixth Edition,* "Chapter 5: Organophosphate Pesticides" (2013), available from the USEPA Office of Pesticide Programs.

[149] Patrick T. O'Shaughnessy, "PARACHUTING CATS AND CRUSHED EGGS The Controversy Over the Use of DDT to Control Malaria," *American Journal of Public Health* 98(11), 1940–1948.

[150] Donald Atwood & Claire Paisley-Jones, "Pesticide Industry Sales and Usage; 2008 - 2012 Market Estimates," Biological and Economic Analysis Division - US EPA Office of Pesticide Programs (2017).

[151] Donald Atwood & Claire Paisley-Jones, "Pesticide Industry Sales and Usage; 2008 - 2012 Market Estimates," Biological and Economic Analysis Division - US EPA Office of Pesticide Programs (2017) and James R Roberts & J Routt Reigart, *Recognition and Management of Pesticide Poisonings: Sixth Edition,* "Chapter 5: Organophosphate Pesticides" (2013) available from the USEPA Office of Pesticide Programs.

[152] Freya Kamel & Jane A. Hoppin, "Association of Pesticide Exposure with Neurologic Dysfunction and Disease," *Environmental Health Perspectives* 112(9): 950–958 (June 2014).

to organophosphates are known to have seizures.[153] Because many of these symptoms are so general (e.g., headache, fatigue, nausea, general malaise, changes in mood) and because most people are not familiar with toxic chemical poisoning, over 80% of even severe organophosphate pesticide poisoning is not readily recognized in patients when they show up in the emergency room.[154]

---

[153] James R Roberts & J Routt Reigart, *Recognition and Management of Pesticide Poisonings: Sixth Edition*, "Chapter 5: Organophosphate Pesticides" (2013), available from the USEPA Office of Pesticide Programs; see also Freya Kamel & Jane A. Hoppin, "Association of Pesticide Exposure with Neurologic Dysfunction and Disease," *Environmental Health Perspectives* 112(9), 950–958 (June 2014).

[154] James R Roberts & J Routt Reigart, *Recognition and Management of Pesticide Poisonings: Sixth Edition*, "Chapter 5: Organophosphate Pesticides" (2013), available from the USEPA Office of Pesticide Programs; see also Freya Kamel & Jane A. Hoppin, "Association of Pesticide Exposure with Neurologic Dysfunction and Disease," *Environmental Health Perspectives* 112(9), 950–958 (June 2014).

CHAPTER 9: THE LEGACY OF WAR AT HOME

> Well one day the people got wind of a plan:
> There's a big lumber baron out to poison our land.
> Kill off the hardwoods; they ain't worth a damn.
> Harvest the fir, ship it off to Japan.
> And we just want to know.
> Where do you think it all goes?
>
> Now, whether you choose a smart business suit,
> A pair of blue jeans,
> Or a strong logger's boot.
> Water is something that everyone drinks
> And you can't clean it up pouring bleach in your sinks.
> So we just want to know, where does it all go?
>
> —Susan Swift Parker[155]

In the 1970s, there was a resurgent interest in homesteading that drew many young families from across the country to settle in Oregon's wilderness.[156] These communities frequently grew their own food, kept their own livestock, and lived largely off the land. They sold what they did not consume. And a notable many sought out organic certification for their fruits, vege-

---

[155] Susan Swift Parker, Just Stories, "Interview with Carol Van Strum and Susan Swift Parker," Conducted by Laura Nausieda & Lauren Rapp, YouTube (uploaded 19, May 2015), available at https://www.youtube.com/watch?v=zK0fAz3_Xn0.

[156] "Carol van Strum and Susan Swift Interview," Lincoln County Community Rights, Interview by Bill Dalby published in YouTube (uploaded March 16, 2019), available at https://www.youtube.com/watch?v=EY7f_H8lIRY.

tables, and meats to sell in the marketplace.[157] Much of this activity was concentrated around the Siuslaw National Forest. During this same time period, the US Forest Service and local timber companies began spraying herbicides from helicopters in and around the area as part of their program to increase timber yields from Douglas fir trees. Some of these herbicides were sprayed simply to eradicate existing hardwood forests to make room for this commercially lucrative tree species.[158] Spraying commenced regularly, not only over forest lands but over entire communities.

As recounted by Carol Van Strum in her book, *A Bitter Fog: Herbicides and Human Rights*,[159] the aerial spraying program in Oregon has been a plague upon the land. Plants died, livestock died, wild animals died. Surviving plants developed deformed leaves and stunted growth. Animals fell ill with cancer and other problems. Human beings developed fatigue, gastrointestinal disorders, organ failure, and cancer. Children, livestock, and wild animals were born with birth defects—many did not survive. Women were miscarrying year after year and hemorrhaged from their uterus.[160] Children suffered severe nosebleeds and bloody diarrhea. Babies were sometimes born with no brain (a birth defect known as anencephaly) and would die soon after birth.[161] Because of these horrors, some people sold their land, gave up their organic farms, and left the area. Others had nowhere else to go.

These communities began organizing. As the crisis grew, enough people voiced their concern that the government and chemical companies could no longer ignore the problem. Instead of ceasing their activities, these powerful interests responded to the public concern with their signature tactics: deny, delay, and do nothing. The authorities eventually came out and took samples from the water, soil, sick humans, and dead fetuses. They would disappear with these and never get back with their results. Or they would issue false water quality results claiming all was well. At best, they would blame disease on pests, lack of vitamins and lifestyle.[162]

---

[157] Carol van Strum, *A Bitter Fog*, 56-63 (Sierra Club Books 1983) (First edition); *A Bitter Fog* (Jerico Hill Interactive 2014) (Second edition).

[158] "Carol van Strum and Susan Swift Interview," Lincoln County Community Rights, Interview by Bill Dalby published in YouTube (uploaded March 16, 2019), available at https://www.youtube.com/watch?v=EY7f_H8IIRY.

[159] Carol van Strum, *A Bitter Fog* (Sierra Club Books 1983) (First edition); *A Bitter Fog* (Jerico Hill Interactive 2014) (Second edition).

[160] Carol van Strum, *A Bitter Fog*, ch. 11 (Sierra Club Books 1983) (First edition); *A Bitter Fog* (Jerico Hill Interactive 2014) (Second edition).

[161] Carol van Strum, *A Bitter Fog*, 210-212, 223 (Sierra Club Books 1983) (First edition); *A Bitter Fog* (Jerico Hill Interactive 2014) (Second edition).

[162] Carol van Strum, *A Bitter Fog*, 25-30, 92-100, 151-160, 203-213 (Sierra Club Books 1983) (First edition); *A Bitter Fog* (Jerico Hill Interactive 2014) (Second edition).

The chemicals being sprayed by the US Forest Service in Oregon were the same chemical combination as found in Agent Orange. Both the US government and the timber companies eagerly adopted the use of Agent Orange during the Vietnam War as a herbicide for maximizing timber yields on public and private lands. After veterans returned with serious chemical poisoning from Agent Orange, the US Congress banned the use of this chemical cocktail in Vietnam.[163] A huge market glut accumulated for Agent Orange in 1970 as a result, and the manufacturers of these chemicals needed to find an alternative market fast. They found this market domestically within the US Forest Service and the timber industry.[164]

Agent Orange is a combination of two chemicals: 2,4-D and 2,4,5-T. These two were responsible for the horrible birth defects, stillbirths, miscarriages, and other health and environmental problems occurring in Oregon a few decades ago.[165] In addition to their own toxic properties, 2,4,5-T and 2,4D have the propensity to generate more toxic chemicals in the manufacturing process—dioxins and furans.[166] Dioxins and furans will form spontaneously in the manufacturing of 2,4-D and 2,4,5-T (they also spontaneously form when burning plastics and during the manufacturing of certain other toxic chemicals). As with many substances, it only takes trace amounts of dioxins and furans to poison the body.[167]

Each of the chemical companies who produced 2,4,5-T and 2,4D (Dow Chemical, Monsanto, Diamond Shamrock, and Hooker Chemical) attested to their safety. The US Department of Agriculture certified these chemicals as safe. Even the US Centers for Disease Control bent over backwards to vouch for their safety. However, the dioxin produced by these chemicals proved to be one of the "most acutely toxic molecules known to man."[168] Dioxin continues its legacy, severely contaminating soil and waterways in

---

[163] Institute of Medicine, *Veterans and Agent Orange: Health Effects of Herbicides Used in Vietnam*, "Chapter 2: History of the Controversy Over the Use of Herbicides," (The National Academies Press 1994).

[164] "Carol van Strum and Susan Swift Interview," Lincoln County Community Rights, Interview by Bill Dalby published in YouTube (uploaded March 16, 2019), available at https://www.youtube.com/watch?v=EY7f_H8IIRY.

[165] E.G. Vallianatos with M. Jenkins, *Poison Spring*, 45 (Bloomsbury Press 2014); Carol van Strum, *A Bitter Fog*, (Jerico Hill Interactive 2014).

[166] James R Roberts & J Routt Reigart, *Recognition and Management of Pesticide Poisonings: Sixth Edition*, "Chapter 10: Chlorophenoxy Herbicides" (2013), available from the USEPA Office of Pesticide Programs; Letter from Monsanto Industrial Chemicals to Diamond Shamrock Agricultural Chemicals and attachments, dated January 18, 1982, available at toxicdocs.org.

[167] US EPA Fact Sheet, "Use of Dioxin TEFs in Calculating Dioxin TEQs at CERCLA and RCRA Sites," (May 2013), available at https://semspub.epa.gov/work/HQ/174558.pdf.

[168] E.G. Vallianatos with M. Jenkins, *Poison Spring*, 46, 69 (Bloomsbury Press 2014).

parts of Virginia and Michigan where Monsanto and Dow Chemical, respectively, produced the main ingredients of Agent Orange for many decades.[169]

Dow Chemical held the main herbicide contract with the US Forest Service in Oregon. The company made a local appearance with its legal staff and PR team to appease the communities. In these meetings, DOW consistently denied that Agent Orange was toxic.[170] The company also pulled strings with the Audubon Society and pressured the US EPA to protect its position. It further engaged in a PR campaign to discredit local residents in the eyes of the nation by claiming they were just some hippies trying to protect their illegal marijuana crop from herbicides.[171] All the while, Dow Chemical knew Agent Orange was deadly.[172]

At the same time, the US Forest Service was conducting research studies on small animals along roadsides sprayed with Agent Orange, finding evidence of bioaccumulation of dioxins in their bodies. The US Forest Service hid from the public the existence of these studies for some time.[173] Only after a federal judge forced the EPA to assess the health impacts of Agent Orange in Oregon did this and other studies come out of the woodwork.[174]

Oregonians were persistent. As a result of public pressure, one of the chemicals used in the aerial spraying formula (2,4,5-T) was withdrawn—it was, however, replaced with other problematic substances (including Roundup, among others).[175] The other ingredient used in Agent Orange, 2,4-D, continues to be in use, even though it was found to be as toxic as 2,4,5,-T.[176] And aerial spraying of toxic pesticides continues in Oregon—some forty years later.[177]

---

[169] E.G. Vallianatos with M. Jenkins, *Poison Spring,* 70-72 (Bloomsbury Press 2014).

[170] Carol van Strum, *A Bitter Fog,* 15-17, 46, 69, 95 (Sierra Club Books 1983) (First edition); *A Bitter Fog* (Jerico Hill Interactive 2014) (Second edition).

[171] Carol van Strum, *A Bitter Fog,* 9, 14, 65, 69, 71, 99, 161, 168, 171, 199, 221 (Sierra Club Books 1983) (First edition); *A Bitter Fog* (Jerico Hill Interactive 2014) (Second edition).

[172] Carol van Strum, *A Bitter Fog,* 133, 142-144, 206, 246-250 (Sierra Club Books 1983) (First edition); *A Bitter Fog* (Jerico Hill Interactive 2014) (Second edition).

[173] "Carol van Strum and Susan Swift Interview," Lincoln County Community Rights, published in YouTube (uploaded March 16, 2019), available at https://www.youtube.com/watch?v=EY7f_H8lIRY.

[174] "Carol van Strum and Susan Swift Interview," Lincoln County Community Rights, published in YouTube (uploaded March 16, 2019), available at https://www.youtube.com/watch?v=EY7f_H8lIRY.

[175] Carol van Strum, *A Bitter Fog,* 60, 209, 216 (Sierra Club Books 1983) (First edition); *A Bitter Fog* (Jerico Hill Interactive 2014) (Second edition); see also Carey Gillam, *Whitewash,* ch. 8 (Island Press 2017).

[176] Carol van Strum, A *Bitter Fog,* 16, 44, 67-75, 217-225 (Sierra Club Books 1983) (First edition); *A Bitter Fog* (Jerico Hill Interactive 2014) (Second edition).

[177] Jes Burns, "Oregon Aerial Pesticide Bills Get Hearings In Salem," *Oregon Public Broadcasting* (April 3, 2019).

While 2,4,5-T was largely abandoned after the 1970s, it is still found in the urine of most Americans.[178] The use of 2,4,D, on the other hand, is steadily increasing.[179] Oregon communities ran out of money and energy to effectively fight the use of 2,4-D for aerial spraying and the EPA has continued to bless the chemical for commercial use. According to 2012 data, 2,4-D is the most common insecticide used in American homes and the fifth most common used by the commercial agricultural sector.[180] Initial symptoms of exposure to 2,4-D can include headache, nausea, abdominal pain, confusion, aggression, muscle weakness, neuropathy, fever, and loss of reflexes.[181]

Many other agricultural communities around the country have suffered from outbreaks of pesticide poisoning as well. In the 1990s, babies were born without brains in the Lower Rio Grande Valley of Southern Texas.[182] There were also incidences of microcephaly in Texas, a birth defect where the brain is not missing, but it is significantly smaller than a normal human brain.[183] The agricultural communities of Yakima Valley in Washington State suffered similar fates during the 2000s—babies were born without brains and others were born with spina bifida, a birth defect in the spine and spinal cord.[184] Spina bifida is also common in the children of Vietnam War Veterans.[185] Regulatory authorities came out to investigate the birth defects in Washington State. They took years to look at the problem, issued lengthy and mostly irrelevant technical reports, and made no substantive conclu-

---

[178] "Fourth National Report on Human Exposure to Environmental Chemicals," US Centers for Disease Control (2019).

[179] Charles Benbrook, "Trends in glyphosate herbicide use in the United States and globally," *Environmental Sciences Europe* 28(3), (February 2016); Carey Gillam, *Whitewash,* 197-202 (Island Press 2017).

[180] Donald Atwood & Claire Paisley-Jones, "Pesticide Industry Sales and Usage; 2008 - 2012 Market Estimates," Biological and Economic Analysis Division - US EPA Office of Pesticide Programs (2017).

[181] James R Roberts & J Routt Reigart, *Recognition and Management of Pesticide Poisonings: Sixth Edition,* "Chapter 10: Chlorophenoxy Herbicides" (2013) available from the USEPA Office of Pesticide Programs.

[182] Dan Oko, "The Toxic Border: Researchers Ponder Increasing Birth Defects in the Valley," *The Austin Chronicle* (Nov. 9, 2011); John McClintock, "Cluster of babies in Texas born without brains Pollution suspected as cause of defect," *The Baltimore Sun* (January 19, 1992).

[183] Hoyt AT, et al., "Pre-Zika Descriptive Epidemiology of Microcephaly in Texas, 2008-2012," *Birth Defects Research* 110(5), 395-405 (March 2018).

[184] James Cheng, "Bizarre Cluster of Severe Birth Defects Haunts Health Experts," *NBC News* (Feb 7, 2014); Jennie McLaurin, "Birth Defects: Anencephaly and Health Justice in Washington State," *Migrant Clinicians Network* (January 20, 2015), available at https://www.migrantclinician.org/blog/2015/jan/birth-defects-anencephaly-and-health-justice-washington-state.html.

[185] Rep. Barbara Lee, H.R. 326, "Victims of Agent Orange Relief Act of 2019," 116th Congress, 1st Session (2019-2020).

sions beyond encouraging pregnant women to take supplements.[186] As a result, these devastating incidences of pesticide poisoning have fallen into public obscurity.

Hawaii has been the most recent victim of pesticide-related birth defects in the country, including children born with severe heart malformations.[187] Hawaii has become an unfortunate target of the pesticide industry as a convenient location for "open air testing of experimental pesticides and GMOs."[188] Companies such as Bayer-Monsanto, Dow-DuPont, Syngenta-ChemChina, and BASF own 13,500 acres in Kauai alone.[189] The pesticides sprayed on these islands include highly toxic chemicals such as chlorpyrifos, atrazine, paraquat, bifenthrin, and glyphosate.[190] These pesticides have drifted into local communities and contaminated their water supplies. Efforts by the Hawaiians to defend themselves and the heavy influence of the pesticide industry on local politicians are featured in a recent documentary called, "Poisoning Paradise."[191]

So what is our federal government doing to protect us from pesticides? Not much. The US has a poor track record on pesticide regulation. A recent study concluded that the US EPA has a history of not banning any pesticide unless the manufacturer freely consents to the ban.[192] Meaning that the US does not take any regulatory action at all—rather, it relies on manufacturers to voluntarily take pesticides off of the market.

The study further concluded that the USA lags in pesticide regulation compared to the EU, Brazil, and China—which are the world's other major pesticide users. Indeed, 72 pesticides that are banned in the European Union for their toxicity are legally in use in the USA, accounting for approximately

---

[186] Washington State Department of Health, "Neural Tube Defect Investigation in Benton, Franklin and Yakima Counties, 2010-2016" (Sept. 2017), available at https://www.doh.wa.gov/Portals/1/Documents/Pubs/210-092-NTDReport.pdf; University of Texas Science Center, "Guide to the Texas Neural Tube Defect Project Archives," available at https://legacy.lib.utexas.edu/taro/uthscsa/00011/hscsa-00011.html (last visited July 23, 2019).

[187] Gary Hoover, "Pesticides & birth defects: Who do you believe?," *The Garden Island* (Nov. 29, 2017) and Christopher Pala, "Pesticides in paradise: Hawaii's spike in birth defects puts focus on GM crops," *The Guardian* (Aug. 23, 2015).

[188] Maggie Sergio, "GMO & Pesticide Experiments in Hawaii: The Poisoning of Paradise," *Huffington Post* (July 2, 2013).

[189] Carey Gillam, *Whitewash*, 135-138 (Island Press 2017).

[190] Carey Gillam, *Whitewash*, 136-139 (Island Press 2017).

[191] "Poisoning Paradise" a documentary by Pierce and Keelly Brosnan (2018), available at https://www.poisoningparadise.com; see also Jimy Tallal, "'Poisoning Paradise' Paints Portrait of Pollution," *The Malibu Times* (Jan. 20, 2018).

[192] Nathan Donley, "The USA lags behind other agricultural nations in banning harmful pesticides," *Environmental Health* 18(44) (June 2019).

322 million pounds of internationally banned toxic pesticides applied to US soil in 2016 alone.[193] The fox is guarding the henhouse.

---

[193] Nathan Donley, "The USA lags behind other agricultural nations in banning harmful pesticides," *Environmental Health* 18(44) (June 2019).

CHAPTER 10: MODERN AGRICULTURE

Modern agriculture is dominated by glyphosate. Over 3.5 billion pounds of glyphosate have been applied in the US since 1974; nearly 19 billion pounds have been applied globally.[194] The majority of glyphosate use has occurred in the last decade and we are using more glyphosate than ever before.[195] The astounding growth rate in the use of this chemical can be attributed to two important factors. First, the introduction of genetically-engineered corn, soy, cotton and other crops that are designed to be glyphosate tolerant. The second is the adoption of what are known as "no-till" farming practices, whereby farmers spray wheat with glyphosate before harvesting to kill off the plant and extract the grain.[196] Between these two practices, farmers have begun to apply glyphosate at unprecedented rates, saturating American farmland. Use has increased fifteen-fold since 1996.[197]

---

[194] Original figures are in kilograms. Global Chemicals Outlook II 104, United Nations Environment Programme (2019) and Charles Benbrook, "Trends in glyphosate herbicide use in the United States and globally," *Environmental Sciences Europe* 28, 3 (February 2016).

[195] Charles Benbrook, "Trends in glyphosate herbicide use in the United States and globally," *Environmental Sciences Europe* 28, 3 (February 2016); see also Donald Atwood & Claire Paisley-Jones, "Pesticide Industry Sales and Usage; 2008 - 2012 Market Estimates," Biological and Economic Analysis Division - US EPA Office of Pesticide Programs (2017).

[196] Charles Benbrook, "Impacts of genetically engineered crops on pesticide use in the U.S. -- the first sixteen years," *Environmental Sciences Europe* 24, 24 (December 2012); see also the documentary by Jeffrey Smith, "Genetic Roulette: The Gamble of Our Lives," (August 2012), available on YouTube at https://www.youtube.com/watch?v=7sUNxX0OxP8.

[197] Charles Benbrook, "Trends in glyphosate herbicide use in the United States and globally," *Environmental Sciences Europe* 28, 3 (February 2016); see also Donald Atwood & Claire Paisley-Jones, "Pesticide Industry Sales and Usage; 2008 - 2012 Market Estimates," Biological and Economic Analysis Division - US EPA Office of Pesticide Programs (2017).

Bayer-Monsanto is the primary producer of glyphosate-based products.[198] RoundUp is the most famous glyphosate-based weed killer, produced and marketed by the company. Monsanto began commercially producing RoundUp for agricultural use in the 1970s. Monsanto obtained the authority to sell glyphosate in the United States by submitting to the EPA lab tests assuring the chemical's safety. These lab tests were performed by IBT Laboratories and were subsequently found to be fraudulent and fake.[199] A federal criminal investigation ultimately shut down the IBT lab for cooking up the majority of its lab tests. Nevertheless, Monsanto and the EPA never looked back at the fraudulent studies. Glyphosate freely took over the marketplace.

Like all pesticides (a category that includes herbicides, insecticides, and biocides), RoundUp is designed to kill life. It is marketed as a product that kills weeds in particular on contact. People wondered whether a product that would cause a plant to die upon contact might not have harmful effects on humans. Monsanto claimed it would not. Instead, Monsanto marketed Roundup and glyphosate as the safe pesticide and sold it to farmers and homeowners worldwide. However, RoundUp and glyphosate proved not to be safe at all.[200]

Monsanto told the public that glyphosate kills weeds through a certain metabolic pathway that exists only in plants—it is called the shikimate pathway. When sprayed on a plant, glyphosate causes a plant to die by poisoning it through this pathway. The human metabolism does not have a shikimate pathway, but the beneficial bacteria living in our gut and other parts of our body do. As a result, they can directly be poisoned by glyphosate when we eat tainted foods and drink contaminated water.[201]

This is a big deal. There are ten-fold more bacteria in our body than actual human cells, and their job is to act as a first line of defense to keep

---

[198] Additionally, "[s]ome two-dozen Chinese firms now supply 40% of the glyphosate used worldwide, and export most of their annual production." Charles Benbrook, "Trends in glyphosate herbicide use in the United States and globally," *Environmental Sciences Europe* 28, 3 (February 2016).

[199] Health Protection Branch, Health and Welfare Canada, Ottawa. "Current recommendations in IBT pesticides:" Press packet, October 19, 1981, *cited in* Carol Van Strum, *A Bitter Fog*, Sierra Club Books (1983) (First edition); *A Bitter Fog* (Jerico Hill Interactive 2014) (Second edition). See also Evaggelos Vallianatos, "The Swamp," *Huffington Post* (August 4, 2017), citing Monsanto Internal Corporate Email from William Heidens to Josh Monken, "RE: CE Collaboration Project," dated March 17, 2015, available at https://usrtk.org.

[200] John Peterson Myers, et al., "Concerns over use of glyphosate-based herbicides and risks associated with exposures: a consensus statement," *Environmental Health* 15,9 (Feb. 2016).

[201] Anthony Samsel and Stephanie Seneff, "Glyphosate, pathways to modern diseases II: Celiac sprue and gluten intolerance," *Interdisciplinary Toxicology* 6(4), 159–184 (Dec. 2013).

us healthy.[202] Our gut microbiome, for example, is involved in helping us produce energy, supports the work of our immune system, and regulates the life and death of our cells.[203] Our gut microbiome contains 150-fold more genetic information than the remainder of our body.[204] This genetic information interacts with our own cells, keeping us functional. Indeed, patients with diabetes, obesity, cardiovascular diseases, allergies, and inflammatory bowel disease, among others, have been found to have an unhealthy gut microbiome.[205] The antimicrobial nature of glyphosate is disruptive to our gut microbiome and helps explain the rising prevalence of gastro-intestinal disorders, food intolerances and serious digestive illnesses such as celiac disease.[206] Intestinal diseases are already well-documented in livestock with the use of glyphosate.[207]

The antimicrobial effects of glyphosate can also kill off essential microbes in the soil. This is a form of soil depletion, and actually causes the plants to become weaker, less productive, and susceptible to devastating disease. Monsanto has gone to great lengths to deny this by attacking scientists who have investigated the antimicrobial effects of glyphosate on agricultural land.[208] At the same time, Monsanto has been well aware of the antimicrobial properties of glyphosate, as it specifically obtained a patent on glyphosate's antimicrobial effects in 2003.[209]

Glyphosate is also in a class of chemicals that disrupts the body's ability to detoxify.[210] This means that we're not able to process and eliminate toxic substances from our body as effectively, and foreign substances can start

---

[202] Robert Naviaux, "Metabolic Features of Cell Danger Response," *Mitochondrion* 16, 7–17 (May 2014).

[203] Kun Lu, et al., "Xenobiotics: Interaction with the Intestinal Microflora," *ILAR Journal* 56(2), 218–227 (Aug. 31, 2015).

[204] Robert Naviaux, "Metabolic Features of Cell Danger Response," *Mitochondrion* 16, 7–17 (May 2014).

[205] Kun Lu, et al., "Xenobiotics: Interaction with the Intestinal Microflora," *ILAR Journal* 56(2), 218–227 (Aug. 31, 2015).

[206] Anthony Samsel and Stephanie Seneff, "Glyphosate, pathways to modern diseases II: Celiac sprue and gluten intolerance," *Interdisciplinary Toxicology* 6(4), 159–184 (Dec. 2013).

[207] Memo from Martin Tang Sørensen, et al., Aarhus University Danish Centre for Food and Agriculture to Danish Ministry of Food, Agriculture and Fisheries, entitled "The feeding of genetically modified glyphosate resistant soy products to livestock" (February 4, 2014), available at https://dca.au.dk/fileadmin/DJF/Notat_gmofoder_uk_version_Memorandum_on_The_feeding_of_genetically_modified_glyphosate_resistant_soy_products_to_livestock.pdf.

[208] Carey Gillam, *Whitewash*, 206-213 (Island Press 2017).

[209] Patent No. US7771736B2, "Glyphosate formulations and their use for the inhibition of 5-enolpyruvylshikimate-3-phosphate synthase" Monsanto Technology LLC (2003).

[210] Anthony Samsel and Stephanie Seneff, "Glyphosate's Suppression of Cytochrome P450 Enzymes and Amino Acid Biosynthesis by the Gut Microbiome: Pathways to Modern Diseases," *Entropy* 15(4), 1416-1463 (April 2013).

to accumulate in our body much faster. Over time, this build-up of toxins can cause a whole host of chronic illnesses, such as inflammatory disorders, obesity, diabetes, heart disease, kidney disease, depression, autism, infertility, cancer and Alzheimer's disease.[211] Unsurprisingly, glyphosate has been found in the urine of chronically ill individuals at markedly higher levels than in healthy individuals.[212]

That is not all. Glyphosate also disrupts essential proteins in the body.[213] Proteins do the legwork in maintaining tissues and organs.[214] They are also responsible for distributing hormones throughout the body to control reproduction, stress responses, brain function, growth, and energy metabolism.[215] Glyphosate's disruption of these proteins is highly problematic; it can ultimate lead to diseases such as "diabetes, obesity, asthma, chronic obstructive pulmonary disease (COPD), pulmonary edema, adrenal insufficiency, hypothyroidism, Alzheimer's disease, amyotrophic lateral sclerosis (ALS), Parkinson's disease, prion (i.e. degenerative) diseases, lupus, mitochondrial disease, non-Hodgkin's lymphoma, neural tube defects, infertility, hypertension, glaucoma, osteoporosis, fatty liver disease and kidney failure."[216]

Researchers have also found a "very strong and highly significant" correlation between increasing use of glyphosate and thyroid, bladder, kidney, pancreatic, and liver cancer; as well as diabetes, inflammatory bowel disease, intestinal disease, obesity, Alzheimer's disease, multiple sclerosis, dementia, and autism, among others.[217]

---

[211] Anthony Samsel and Stephanie Seneff, "Glyphosate's Suppression of Cytochrome P450 Enzymes and Amino Acid Biosynthesis by the Gut Microbiome: Pathways to Modern Diseases," *Entropy* 15(4), 1416-1463 (April 2013); Anthony Samsel and Stephanie Seneff, "Glyphosate, pathways to modern diseases II: Celiac sprue and gluten intolerance," *Interdisciplinary Toxicology* 6(4), 159–184 (December 2013).

[212] Monika Krüger, et al., "Detection of Glyphosate Residues in Animals and Humans," *Environmental & Analytical Toxicology* 4:2 (2014).

[213] Anthony Samsel and Stephanie Seneff, "Glyphosate pathways to modern diseases V: Amino acid analogue of glycine in diverse proteins," *Journal of Biological Physics and Chemistry* 16, 9 – 46 (March 2016); and Anthony Samsel, PPT Presentation, "Glyphosate Herbicide Pathways To Modern Diseases Synthetic Amino Acid And Analogue of Glycine Mis-incorporation Into Diverse Proteins," (Dec. 2016), available at https://people.csail. mit.edu/seneff/DC2016/AnthonySamsel_DC2016.pdf.

[214] National Institutes of Health US National Library of Medicine, "Genetics Home Reference: What are proteins and what do they do?," available at https://ghr.nlm.nih. gov/primer/howgeneswork/protein (last visited July 17, 2019).

[215] European Union, "Endocrine Disruptors: from Scientific Evidence to Human Health Protection," a study commissioned by the PETI Committee of the European Parliament 18 (2019).

[216] Anthony Samsel and Stephanie Seneff, "Glyphosate pathways to modern diseases V: Amino acid analogue of glycine in diverse proteins," *Journal of Biological Physics and Chemistry* 16, 9–46 (March 2016).

[217] Swanson, N.L., Leu, A., Abrahamson, J. & Wallet, B. "Genetically engineered crops, glyphosate and the deterioration of health in the United States of America," *Journal of*

Monsanto has known for many years that glyphosate is toxic to and accumulated in the organs, digestive system, fat, muscle, bone marrow, glands, and blood of the human body.[218] Monsanto's own researchers found that glyphosate causes tissue destruction to the pituitary, thyroid, thalamus, testes, adrenal glands, and the major organs. Monsanto researchers also found that glyphosate causes DNA damage in bone marrow, pancreatic tumors; cataracts; kidney disease; and serious lung problems.[219] Monsanto's research studies were obtained from the EPA by scientists under the Freedom of Information Act.[220] Despite the fact that the EPA was in possession of these research studies for quite some time, the EPA protected Monsanto while doing nothing to protect human health from glyphosate.[221]

If this were not bad enough, time has revealed that a surfactant used in the chemical formulation of RoundUp is also toxic to human cells and very effective at getting into them. This surfactant, when combined with Round-Up's main ingredient glyphosate, has been associated with non-Hodgkins lymphoma, a kind of cancer that is more prevalent for those who apply pesticides professionally. There is also a strong correlation between glyphosate and "breast cancer, pancreatic cancer, kidney cancer, thyroid cancer, liver cancer, bladder cancer and myeloid leukemia."[222]

The literature linking pesticides to cancer in farm workers was already voluminous in the US by the late 1980s.[223] Although a diversity of cancers has been associated with pesticides, lymphoma was probably the most prevalent cancer found in these research studies, in vastly higher rates than liver,

Organic Systems 9(2) 6–37 (January 2014); see also Carey Gillam, *Whitewash*, ch. 5 (Island Press 2017).

[218] Anthony Samsel, PPT Presentation, "Glyphosate Herbicide Pathways To Modern Diseases Synthetic Amino Acid And Analogue of Glycine Mis-incorporation Into Diverse Proteins," (Dec. 2016), available at https://people.csail.mit.edu/seneff/DC2016/AnthonySamsel_DC2016.pdf.

[219] Anthony Samsel and Stephanie Seneff, "Glyphosate pathways to modern diseases V: Amino acid analogue of glycine in diverse proteins," *Journal of Biological Physics and Chemistry* 16, 9–46 (March 2016) (citing internal Monsanto research studies).

[220] Anthony Samsel and Stephanie Seneff, "Glyphosate, pathways to modern diseases IV: cancer and related pathologies," *Journal of Biological Physics and Chemistry*, 15(3), 121-159 (January 2015).

[221] Carey Gillam, a reporter who covers the agrochemical industry, writes extensively about the EPA's involvement in Monsanto's cover-up of the health effects of glyphosate in her book. See Carey Gillam, *Whitewash* (Island Press 2017). More information on Monsanto's influence over our government can be found in the documentary, "The World According to Monsanto." Marie-Monique Robin, "The World According to Monsanto" (2008).

[222] Anthony Samsel and Stephanie Seneff, "Glyphosate, pathways to modern diseases IV: cancer and related pathologies," *Journal of Biological Physics and Chemistry* 15, 121–159 (August 2015).

[223] Marion Moses, "Cancer in Humans and Potential Occupational and Environmental Exposure to Pesticides: Selected Epidemiological Studies and Case Reports," *AAOHN Journal* 37:3, 131-136 (March 1989).

stomach, prostate, skin, and brain cancer, which also crop up frequently in the research. Glyphosate was linked to non-Hodgkins lymphoma as early as 2000.[224] There is a whole body of literature since 2000 linking glyphosate to non-Hodgkins lymphoma.[225]

Non-Hodgkin's lymphoma sickened a California groundskeeper who used RoundUp for work. DeWayne Johnson, age 46 and destined to die an early and very painful death, refused to settle litigation with Monsanto over the disease. Johnson was determined to take the company to trial so that the public would get to see what the company has been up to all these years. The litigation revealed that Monsanto publicly lied about the safety of Roundup to consumers and colluded with the EPA to cover up the evidence.[226] Monsanto also ghost-wrote scientific papers falsely touting the safety of RoundUp, harassed opposing scientists, and worked to mislead regulatory agencies around the world.[227] The company further hired internet trolls to manipulate information about RoundUp on social media and digital news outlets.[228] The judge who reviewed Mr. Johnson's case before trial had the following to say about the evidence:

> Monsanto has long been aware of the risk that its glyphosate-based herbicides are carcinogenic...but has continuously sought to influence the scientific literature to prevent its internal concerns from reaching the public....[229]

People are waking up to the dangers of glyphosate, although the EPA has yet to take reasonable action. Indeed, as of the writing of this book, the

---

[224] Lennart Hardell & Mikael Eriksson, "A Case Control Study of Non-Hodgkin Lymphoma and Exposure to Pesticides," *American Cancer Society Journal of Cancer* 85(6), 1353-1360 (March 15, 1999).

[225] Luoping Zhang, et al., "Exposure to Glyphosate-Based Herbicides and Risk for Non-Hodgkin Lymphoma: A Meta-Analysis and Supporting Evidence" *Mutation Research / Reviews in Mutation Research* 781, 186-206 (July-Sept, 2019).

[226] Baum Hedlund Aristei & Goldman website, "Monsanto Papers | Secret Documents," available at: https://www.baumhedlundlaw.com/toxic-tort-law/monsanto-roundup-lawsuit/monsanto-secret-documents/ (last visited on May 14, 2019); and Baum Hedlund Aristei & Goldman, "The Monsanto Papers — Master Chart," (updated April 24, 2019) available at https://www.baumhedlundlaw.com/pdf/monsanto-documents-2/Monsanto-Papers-Master-Chart.pdf.

[227] Leemon McHenry, "The Monsanto Papers: Poisoning the Scientific Well" *International Journal of Risk, Safety & Medicine* 29(3-4), 193-205 (June 2018).

[228] In re Roundup Products Liability Litigation, Case No. 16-md-02741-VC, "Plaintiffs' Reply in Support of Motion to Strike Confidentiality of Heydens Deposition," US District Court, Northern District of California (April 20, 2017).

[229] "Judicial Order on (1) Monsanto's Omnibus Sargon Motion; (2) Monsanto's Motion for Summary Judgment; (3) Plaintiff's Omnibus Sargon Motion; (4) Plaintiff's Motion for Summary Adjudication," 45, Dewayne Johnson v. Monsanto, Case No. CGC - 16-550128, Superior Court of California - County of San Francisco (May 16, 2018).

EPA continues to claim that glyphosate is safe.[230] Moreover, much damage has been done. Glyphosate has been detected ubiquitously in our air, rain, groundwater, surface water, soil, and even seawater.[231] Glyphosate has also contaminated the majority of our food supply.[232] The chemical is migrating into unborn generations: over 90% of pregnant women in the Midwest and 86% of pregnant women in other parts of the US have tested positive for glyphosate.[233]

Glyphosate has not only been a plague upon human health, but also a plague upon agriculture. RoundUp has been responsible for an alarming increase in the growth of super weeds on agricultural land.[234] These are weeds that can grow up to 3 inches a day, become 8 feet tall, and are stout enough to damage farm machinery. Farmers generally have to hire workers to pull these weeds by hand in order to stop them from destroying their crops completely. These weeds have spread quickly in just a short time, contaminating to over 70 million acres or nearly 18% of US farmland.[235]

At the same time, RoundUp continues to be aggressively marketed to farmers and homeowners. In the spring and summer of 2019, RoundUp and similar weed killers were prominently displayed in the most front-facing aisles at Lowes and Home Depot stores; no doubt Lowes and Home Depot were paid a hefty amount by Bayer-Monsanto for this prominent display of RoundUp. In addition, RoundUp is still being sprayed for its weed-killing properties by state, local, and/or private parks management staff in even some of the most pristine and remote natural areas.[236]

In 2016, Monsanto was bought out by Bayer, an even larger company with ambitions to maintain and secure its market share in the global commercial

---

[230] "US EPA Continues Glyphosate Cancer Cover Up with Regulatory Review Publication," *Sustainable Pulse* (January 31, 2020).

[231] Swanson, N.L., et al., "Genetically engineered crops, glyphosate and the deterioration of health in the United States of America." *Journal of Organic Systems* 9(2), 6–37 (Jan. 2014).

[232] Roy Gerona and Axel Adams, "Glyphosate Biomonitoring: Challenges and Opportunities," Presentation at the Biomonitoring California Scientific Guidance Panel Meeting (July 20, 2017), available at https://biomonitoring.ca.gov/sites/default/files/downloads/GeronaAdams07202017.pdf; Carey Gillam, *Whitewash*, 55-77 & ch 9 (Island Press 2017).

[233] Roy Gerona and Axel Adams, "Glyphosate Biomonitoring: Challenges and Opportunities" Presentation at the Biomonitoring California Scientific Guidance Panel Meeting (July 20, 2017), available at https://biomonitoring.ca.gov/sites/default/files/downloads/GeronaAdams07202017.pdf.

[234] See Carey Gillam, *Whitewash*, 189-200 (Island Press 2017).

[235] Carey Gillam, *Whitewash*, 192 (Island Press 2017) cf. Daniel P. Bigelow and Allison Borchers, "Major Uses of Land in the United States, 2012," United States Department of Agriculture (2017).

[236] I observed a notice posted in May 2019 that Roundup had been sprayed to control weeds in the remote and pristine region of the Cascade Lakes in Central Oregon.

production of biocides such as RoundUp. Bayer is a major player in the pharmaceuticals industry. The company is strategically positioning itself to sell us pharmaceuticals when we become sick from exposure to chemicals such as glyphosate. That means the sicker we get on Bayer-Monsanto's pesticides and GMO crops, the more money Bayer-Monsanto makes on selling us drugs to make us feel better. Thus, the company is beautifully poised in the marketplace to provide us both the poison and the cure. A great business proposition for them; a terrible health and economic loss for the rest of us. Dow Pharmaceutical Sciences and others are undoubtedly banking on the same type of market windfall.

Just like DDT, the herbicides used in Agent Orange, and others, glyphosate may eventually fall out of favor for commercial use after decades of deceit and citizen efforts to protect themselves. Nevertheless, we will feel the brunt of the damage for a long time. In addition, it is highly likely that glyphosate will be replaced by another regrettable substitute pushed by the pesticide industry. Chemical companies have been busy cultivating the next generation of pesticides and pesticide-resistant GMO crops to secure their growing market share. 2,4-D (a key ingredient in Agent Orange) and dicamba are both potential successors to glyphosate.[237]

Farmers not associated with big agriculture want to farm very differently—without the use of pesticides, genetically modified crops, or patented seeds. Farming has successfully been managed without pesticides for millennia in a way that is safe and humane to people, livestock, crops, and the natural world. This type of farming is known as agroecology.[238] Unlike what big agriculture has to offer, agroecology is sustainable, utilizing biodiversity to keep pests in check, promoting food abundance and establishing economic resilience. The agrochemical industry is not interested, however.

---

[237] Charles Benbrook, "Trends in glyphosate herbicide use in the United States and globally," *Environmental Sciences Europe* 28, 3 (February 2016); Carey Gillam, *Whitewash*, 197-202 (Island Press 2017).

[238] Marcia Ishii-Eiteman, "Global Groundswell for Agroecology," *The Catalyst, Pesticide Action Network North America* (2019). For a documentary demonstrating how species diversity promotes agricultural yields and farm resiliency, see John Chester & Sandra Keats, "The Biggest Little Farm" (May 2019), available at https://www.biggestlittlefarmmovie.com.

Chapter 11: Forever Chemicals

> [The] devil doesn't come dressed in a red cape and pointy horns.
> He comes as everything you've ever wished for....
>> —Tucker Max[239]

The year 2019 was filled with news reports regarding contamination of US waterways with a class of forever chemicals known as PFAS. PFAS are an astoundingly large group of chemicals, numbering nearly five thousand separate compounds.[240] They go by countless names, such as PFOA, PTFE, PFOS, PFCs, CFCs, GenX, C8, and many more. Sometimes multiple names are given to a single chemical. Other times one name refers to a subset of these chemicals. And in yet other cases, one name can refer to a single chemical and multiple chemicals at the same time.

For example, C8 is a synonym for PFOA—a single PFAS chemical. C8 can also refer to certain PFAS chemicals that were in DuPont's manufacturing portfolio and which have eight carbon molecules.[241] All of these synonyms

---

[239] Tucker Max, *Assholes Finish First* (Simon & Schuster 2010).

[240] OECD Environment, Health and Safety Publications Series on Risk Management No. 39 "Toward a New Comprehensive Global Database of Per- and Polyfluoroalkyl Substances Substances (PFASs): Summary Report on Updating the OECD 2007 List of Per- and Polyfluoroalkyl Substances (PFASs) (May 4, 2018) & 2018 OECD PFASs Spreadsheet, "Toward a New Comprehensive Global Database of Per- and Polyfluoroalkyl Substances (PFASs)" (OECD 2018), available at https://www.oecd.org/chemicalsafety/portal-perfluorinated-chemicals.

[241] Testimony of Daryl Roberts, Chief Operations & Engineering Officer at Dupont de Nemours, Inc. in front of US House of Representatives Committee on Oversight & Reform (116th Congress), during the "The Devil They Knew: PFAS Contamination and the Need for Corporate Accountability, Part II" Hearing (Sept. 10, 2019).

make it very confusing for the public to keep track of PFAS. And it makes it very convenient for industry to keep using them without a whole lot of scrutiny.

PFAS are a class of chemicals that have been around since World War II. Indeed, they were initially developed for military applications.[242] After World War II, 3M began to mass produce them for use in commercial products.[243] Many scientists who received their education before the 1990s— before problems with PFAS started to leak into the public—were taught in school that PFAS were safe because they are chemically very stable. Their chemical stability did not correlate with actual safety, however. Indeed, their chemical stability turned out to be one of their most devilish qualities.

Because they are chemically so stable, PFAS are very hard to get rid of once they get into our body.[244] Nor do they remain inert; rather, they are actually quite active, wreaking havoc. PFAS chemicals are known to be powerful hormone disruptors (Chapters 15–17), whose toxic effects have been shown in humans.[245] In a recent epidemiological study, researchers in the Veneto Region of Italy compared boys who were exposed to long-chain PFAS even before birth to those who had not had PFAS exposure.[246] The exposed boys had on average five times more PFAS in their blood. The researchers also found that the majority of the exposed boys had smaller testicles, smaller penises, low sperm quality, significantly lower semen counts, and a shorter distance between their anus and genitals. Similar types of birth defects had

---

[242] Sharon Lerner, "3M Knew About the Dangers of PFOA and PFOS Decades Ago, Internal Documents Show," *The Intercept* (July 31, 2018).

[243] Testimony of Denise R. Rutherford, Senior Vice President of Corporate Affairs for The 3M Company in front of US House of Representatives Committee on Oversight & Reform (116th Congress), during the "The Devil They Knew: PFAS Contamination and the Need for Corporate Accountability, Part II" Hearing (Sept. 10, 2019).

[244] See e.g., Philippe Grandjean, "PFC Toxicity in Children" (January 9, 2012) available at https://www.youtube.com/watch?v=cOFFQ7cPAOM.

[245] A.C. Gore, et al., "EDC-2: The Endocrine Society's Second Scientific Statement on Endocrine-Disrupting Chemicals," *Endocrine Review* 36(6), E1–E150 (Dec. 2015); "Endocrine Disruptors: from Scientific Evidence to Human Health Protection," a study commissioned by the PETI Committee of the European Parliament (2019); Allan Astrup Jensen & Henrik Leffers, "Emerging endocrine disrupters: perfluoroalkylated substances," *International Journal of Andrology* 31:2, 39 (March 2008); OECD Environment, Health and Safety Publications Series on Risk Management No. 39 "Toward a New Comprehensive Global Database of Per- and Polyfluoroalkyl Substances (PFASs): Summary Report on Updating the OECD 2007 List of Per- and Polyfluoroalkyl Substances (PFASs)" 8 (May 4, 2018); Di Nisio, et al., "Endocrine disruption of androgenic activity by perfluoroalkyl substances: clinical and experimental evidence," *Journal of Clinical Endocrinology & Metabolism* 104(4), 1259-1271 (April 2019).

[246] Di Nisio, et al., "Endocrine disruption of androgenic activity by perfluoroalkyl substances: clinical and experimental evidence." *Journal of Clinical Endocrinology & Metabolism* 104(4), 1259-1271 (April 2019).

---

been documented by other scientists over the decades in areas polluted with hormone disruptors.[247]

PFAS are also strong immune suppressants and are demonstrated to significantly reduce our immune response to vaccines.[248] Moreover, a variety of serious health problems have been linked to PFAS in human beings, including thyroid problems, neurodevelopment problems in children, auto-immune disease, ulcerative colitis, liver toxicity, kidney cancer, testicular cancer, obesity, and high cholesterol.[249]

Despite their toxicity, PFAS are very popular commercially. They are water-repellent and oil-repellent, heat resistant, and flame resistant.[250] As a result, they are broadly used in consumer products.[251] Indeed, PFAS are the foundation of the Gore-Tex, Scotch Gard, Stainmaster, and Polar Tec brands. They are also used as medical implants, surface coatings on furniture and carpeting, flame retardants, and weatherproofing for outdoor gear and apparel (shoes, clothing, tents, etc.). PFAS are further used in food packaging (pizza boxes, popcorn bags, fast-food containers) to create a grease-proof barrier. Some of the most unsuspecting products containing PFAS include

---

[247] Theo Colborn, et al., *Our Stolen Future*, 6, 189 (Penguin Group 1997).

[248] Philippe Grandjean, et al., "Serum Vaccine Antibody Concentrations in Adolescents Exposed to Perfluorinated Compounds," 125:7 (July 2017); A.C. Gore, et al., "EDC-2: The Endocrine Society's Second Scientific Statement on Endocrine-Disrupting Chemicals," *Endocrine Review* 36(6), E1–E150 (Dec. 2015); Endocrine Disruptors: from Scientific Evidence to Human Health Protection," a study commissioned by the PETI Committee of the European Parliament 39 (2019).

[249] A.C. Gore, et al., "EDC-2: The Endocrine Society's Second Scientific Statement on Endocrine-Disrupting Chemicals," *Endocrine Review* 36(6), E1–E150 (Dec. 2015); Endocrine Disruptors: from Scientific Evidence to Human Health Protection," a study commissioned by the PETI Committee of the European Parliament (2019); Blum, A., S.A. Balan, et al., "The Madrid Statement on Poly- and Perfluoroalkyl Substances (PFASs)," *Environmental Health Perspectives* 123(5), A107-11 (May 2015).

[250] OECD Environment, Health and Safety Publications Series on Risk Management No. 39 "Toward a New Comprehensive Global Database of Per- and Polyfluoroalkyl Substances Substances (PFASs): Summary Report on Updating the OECD 2007 List of Per- and Polyfluoroalkyl Substances (PFASs) (May 4, 2018).

[251] OECD Environment, Health and Safety Publications Series on Risk Management No. 39 "Toward a New Comprehensive Global Database of Per- and Polyfluoroalkyl Substances Substances (PFASs): Summary Report on Updating the OECD 2007 List of Per- and Polyfluoroalkyl Substances (PFASs) (May 4, 2018); "Poisoned Legacy: Where Consumers Encounter PFCS Today," *Environmental Working Group* (May 1, 2015), available at https://www.ewg.org/research/poisoned-legacy/where-consumers-encounter-pfcs-today; Elsie Sunderland, et al., "A review of the pathways of human exposure to poly- and perfluoroalkyl substances (PFASs) and present understanding of health effects," *Journal of Exposure Science & Epidemiology* 29, 131-147 (2019); "EWG's Guide to Avoiding PFAS Chemicals: a Family of Chemicals You Don't Want Near Your Family," *Environmental Working Group* (Updated June 2018), available at https://static.ewg.org/ewg-tip-sheets/EWG-AvoidingPFCs.pdf.

some brands of eye shadow and shaving cream; ski wax; the wristband of the Apple watch; and Oral-B Glide dental floss.[252]

Because PFAS do not break down easily, they have been widely found in the blood of human beings and animals all around the world.[253] A subcategory of PFAS, known as "long-chain" PFAS,[254] are chemically so complex and persistent that they never seem to break down at all. They also bio-accumulate, meaning they build up in our bodies and the food chain over time.

Up until recently, the government did not pay attention to PFAS at all. They were legally grandfathered in as presumably safe under the Toxic Substances Control Act (TSCA) during the law's passage.[255] Indeed, the EPA did not even know about their existence for many decades. Recently, the OECD lamented that although more PFAS are being developed and used in commerce over time, regulators have not been able to keep up with "identi-

---

[252] Katherine Boronow, et al., "Serum Concentrations of PFASs and Exposure-Related Behaviors in African American and Non-Hispanic White Women," *Journal of Exposure Science & Epidemiology* 29, 216-217 (2019): Elsie Sunderland, et al., "A review of the pathways of human exposure to poly- and perfluoroalkyl substances (PFASs) and present understanding of health effects," *Journal of Exposure Science & Epidemiology* 29, 131-147 (2019); "EWG's Guide to Avoiding PFAS Chemicals: a Family of Chemicals You Don't Want Near Your Family," Environmental Working Group (Updated June 2018), available at https://static.ewg.org/ewg-tip-sheets/EWG-AvoidingPFCs.pdf.

[253] Blum, A., S.A. Balan, et al., "The Madrid Statement on Poly- and Perfluoroalkyl Substances (PFASs)," *Environmental Health Perspectives* 123(5): p. A107-11 (May 2015); U.S. Environmental Protection Agency Website, "Research on Per- and Polyfluoroalkyl Substances (PFAS)," available at https://www.epa.gov/chemical-research/research-and-polyfluoroalkyl-substances-pfas (last visited on April 25, 2019); "Global Chemicals Outlook II" 50, ISBN No: 978-92-807-3745-5, United Nations Environment Programme (2019); "Toward a New Comprehensive Global Database of Per- and Polyfluoroalkyl Substances Substances (PFASs): Summary Report on Updating the OECD 2007 List of Per- and Polyfluoroalkyl Substances (PFASs)" 8, OECD Environment, Health and Safety Publications Series on Risk Management No. 39 (May 4, 2018); CDC National Biomonitoring Program, "Per- and Polyfluorinated Substances (PFAS) Factsheet," (updated April 7, 2017), available at https://www.cdc.gov/biomonitoring/PFAS_FactSheet.html; Marc A. Mills, "Welcome to the Per- and Polyfluoroalkyl Substances (PFAS) Heartland Community Engagement," EPA Office of Ground Water & Drinking Water Region 7- Leavenworth, Kansas (September 5, 2018), available at https://www.epa.gov/sites/production/files/2018-09/documents/final_epa_pfas_leavenworth_kansas_presentations_september_5_2018.pdf.

[254] The OECD explains that "Based on the commonly accepted OECD definition, long-chain PFAAs refer to perfluoroalkyl carboxylic acids (PFCAs) with ≥ 7 perfluorinated carbons and perfluoroalkane sulfonic acids (PFSAs) with ≥ 6 perfluorinated carbons." "Toward a New Comprehensive Global Database of Per- and Polyfluoroalkyl Substances Substances (PFASs): Summary Report on Updating the OECD 2007 List of Per- and Polyfluoroalkyl Substances (PFASs), OECD Environment, Health and Safety Publications Series on Risk Management No. 39 (May 4, 2018). Note that individual governments and other entities do not necessarily stick to this definition.

[255] "For the First Time in 40 Years EPA to Put in Place a Process to Evaluate Chemicals that May Pose Risk," US EPA (Jan 1, 2017), available at https://archive.epa.gov/epa/newsreleases/first-time-40-years-epa-put-place-process-evaluate-chemicals-may-pose-risk.html.

---

fying and understanding [their] production, use, releases and environmental presence...on the global market..."[256] There is no sign of stopping this trend.

3M was one of the original producers of these chemicals, manufacturing four distinct PFAS in Minnesota for nearly 50 years.[257] The company was aware from the 1950s that they were dangerous. Nevertheless, 3M dumped these chemicals as waste from its production facilities into local waterways, contaminating the drinking supply.[258] 3M also actively influenced the scientific literature around the toxicity of these chemicals in order to protect its business interests. In its own words, 3M set out to "command the science" to maintain its market share.[259] The company selectively funded outside research to cover up the dangers of these chemicals. It reviewed scientific research before it got published to decide whether to fight it. The company went so far as to retain the editor of a number of academic journals for the purpose of screening out potentially unfavorable publications.[260] 3M was very successful at keeping scientific scrutiny at bay and made billions of dollars on these chemicals as a result.[261]

While 3M was making a killing, a number of 3M scientists and attorneys became alarmed by internal corporate studies indicating that PFAS were highly toxic. These employees raised concern about PFAS to 3M's leadership on multiple occasions. 3M was also notified by a third party in 1975 that

---

[256] "Toward a New Comprehensive Global Database of Per- and Polyfluoroalkyl Substances Substances (PFASs): Summary Report on Updating the OECD 2007 List of Per- and Polyfluoroalkyl Substances (PFASs)" 8, OECD Environment, Health and Safety Publications Series on Risk Management No. 39 (May 4, 2018).

[257] State of Minnesota v. 3M Company, "Memorandum in Support of Plaintiff State of Minnesota's Motion to Amend Complaint," Fourth Judicial District Court Hennepin County, Minnesota, File No. 27-CV-10-28862 (Nov. 17, 2017).

[258] State of Minnesota v. 3M Company, "Memorandum in Support of Plaintiff State of Minnesota's Motion to Amend Complaint," Fourth Judicial District Court Hennepin County, Minnesota, File No. 27-CV-10-28862 (Nov. 17, 2017).

[259] Testimony of Lori Swanson, former Attorney General, State of Minnesota in front of US House of Representatives Committee on Oversight & Reform (116th Congress), during the "The Devil They Knew: PFAS Contamination and the Need for Corporate Accountability, Part II" Hearing (Sept. 10, 2019) and State of Minnesota v. 3M Company, "Memorandum in Support of Plaintiff State of Minnesota's Motion to Amend Complaint," Fourth Judicial District Court Hennepin County, Minnesota, File No. 27-CV-10-28862, Nov. 17, 2017.

[260] Testimony of Lori Swanson, former Attorney General, State of Minnesota in front of US House of Representatives Committee on Oversight & Reform (116th Congress), during the "The Devil They Knew: PFAS Contamination and the Need for Corporate Accountability, Part II" Hearing (Sept. 10, 2019).

[261] State of Minnesota v. 3M Company, "Memorandum in Support of Plaintiff State of Minnesota's Motion to Amend Complaint," Fourth Judicial District Court Hennepin County, Minnesota, File No. 27-CV-10-28862 (Nov. 17, 2017).

PFAS were being found in the blood of Americans.[262] The company, however, decided to ignore the evidence.[263]

Things changed when a 3M whistleblower raised public concern about PFAS.[264] This shook up 3M enough that the company "voluntarily" approached the EPA in 2000 and committed to ceasing production of two of its PFAS chemicals.[265] Soon after, 3M found that the American food supply was contaminated with these chemicals—including milk, bread, ground beef, and even apples.[266] This troubled the company even more.

3M's change of heart did not sit well with DuPont. DuPont was a major customer of 3M since the 1950s. Indeed, PFAS were foundational to DuPont, who built an empire using these chemicals to make Teflon and the non-stick frying pans that still decorate the majority of the households in the US (if not the world). Having lost 3M as its major supplier, DuPont decided to produce PFOA (one of the PFAS chemicals that 3M "voluntarily" stopped producing) in order to fuel its frying pan empire.[267]

DuPont had already been notified by 3M in the 1970s that this particular PFAS was persistent, bio-accumulative and toxic. DuPont also had its own corporate studies dating back to the 1950s to verify this.[268] The company had

---

[262] Testimony of Lori Swanson, former Attorney General, State of Minnesota in front of US House of Representatives Committee on Oversight & Reform (116th Congress), during the "The Devil They Knew: PFAS Contamination and the Need for Corporate Accountability, Part II" hearing (Sept. 10, 2019).

[263] Testimony of Rob Bilott, Taft, Stettinus & Hollister LLP in front of US House of Representatives Committee on Oversight & Reform (116th Congress), during the "The Devil They Knew: PFAS Contamination and the Need for Corporate Accountability, Part II" Hearing (Sept. 10, 2019); and Testimony of Lori Swanson, former Attorney General, State of Minnesota in front of US House of Representatives Committee on Oversight & Reform (116th Congress), during the "The Devil They Knew: PFAS Contamination and the Need for Corporate Accountability, Part II" Hearing (Sept. 10, 2019).

[264] State of Minnesota v. 3M Company, "Memorandum in Support of Plaintiff State of Minnesota's Motion to Amend Complaint," Fourth Judicial District Court Hennepin County, Minnesota, File No. 27-CV-10-28862, Nov. 17, 2017, available at http://www.toxicdocs.org.

[265] "EPA and 3M ANNOUNCE PHASE OUT OF PFOS," US EPA (May 16, 2000), available at https://archive.epa.gov/epapages/newsroom_archive/newsreleases/33aa946e6cb11f35 852568e1005246b4.html. See also Letter from 3M to US EPA dated March 13, 2003, re "Environmental Health and Safety Measures Related to Perfluorooctanoic Acid and its Salts (PFOA)," available at http://www.fluoridealert.org/wp-content/pesticides/effect. pfoa.class.mar.13.2003.pdf.

[266] ³M Analytical Report, "Analysis of PFOS, FOSA, and PFOA from Various Food Matrices Using HPLC Electrospray/Mass Spectrometry" (June 21, 2001), attached to Letter from Robert Bilott to US FDA re "PFAS in Food Supply" (June 11, 2019), available at https://docs.house.gov/meetings/GO/GO28/20190910/109902/HHRG-116-GO28-20190910-SD006.pdf.

[267] Leach v. DuPont, "Order on Class Certification and Related Motions," Case No. 01-C-608, Circuit Court of Wood County West Virginia (Jan. 15, 2003), available at toxicdocs.org.

[268] Leach v. DuPont, "Order on Class Certification and Related Motions," 6-9. Case No. 01-C-608, Circuit Court of Wood County West Virginia, (Jan. 15, 2003).

---

found troubling results in its own research on rodents, chimpanzees, and even the company's employees.[269] The desire for growth, however, was much greater for DuPont than its concerns over these studies. DuPont decided to continue use of its highly toxic chemical for a profit, rationalizing that it was "the devil we know."[270]

At this point, you might be wondering why the EPA would not prevent DuPont from manufacturing PFOA, given that 3M had told regulators that the chemical is toxic. That would be a reasonable question to ask. The EPA unfortunately did nothing to intervene. DuPont was able to manufacture PFOA without restriction and continue exporting its frying pans across the world.

To facilitate production, DuPont began dumping wastewater contaminated with PFOA into the Ohio River outside of its manufacturing facility in Parkersburg, West Virginia.[271] It also illegally dumped the chemical into landfill, which also ultimately leached into local waterways.[272] DuPont knew by 1984 that PFOA had accumulated in the local water supply above the company's own safety guidelines, but they did not tell anyone.[273]

The citizens of DuPont's hometown ultimately became suspicious that all was not well because of sudden livestock deaths, unexplained birth defects, and numerous cancers that took the lives of DuPont employees. Earl Tennant, a local farmer who lost an entire herd of cattle to gruesome disease, decided to file suit. He reached out to Robert Bilott, a corporate defense attorney who normally represented chemical companies. Against all odds, Bilott and his law firm took an interest in the case.

The Tennant litigation revealed that DuPont had been dumping into landfill and into the Ohio River what was at the time an unregulated and undisclosed toxic chemical known as PFOA (aka C8).[274] This knowledge, which took years to fully extract out of DuPont, subsequently led to a

---

[269] Robert Bilott, *Exposure* 72–77 (Simon & Schuster 2019).

[270] DuPont Internal Company Document quoted in the documentary by Stephanie Soechtig & Jeremy Siefert, *The Devil We Know* (2018).

[271] Leach v. DuPont, "Order on Class Certification and Related Motions," Case No. 01-C-608, Circuit Court of Wood County West Virginia (Jan. 15, 2003).

[272] See generally Robert Bilott, *Exposure* (Simon & Schuster 2019).

[273] Testimony of Rob Bilott, Taft, Stettinus & Hollister LLP in front of US House of Representatives Committee on Oversight & Reform (116th Congress), during the "The Devil They Knew: PFAS Contamination and the Need for Corporate Accountability, Part II" Hearing (Sept. 10, 2019).

[274] Robert Bilott, *Exposure*, ch. 5 (Simon & Schuster 2019); see also See C8 Science Panel, C8 Study Publications, updated November 2013 (last visited June 10, 2019), available at http://www.c8sciencepanel.org/publications.html; see also Branford Pate, "Diamond Materials for Solvated Electron Chemistry - Potential Payoff: Remediation of AFFF PFAS Contamination from Water" (Nov 30, 2016 revision), available at http://www.toxicdocs.org.

massive class-action lawsuit regarding the company's dumping of PFOA / C8 into the Ohio River.[275]

DuPont's leadership was deposed many times before trial about the dumping of this chemical. Most of them were largely in denial, as illustrated by the following interaction:

> Question from Plaintiffs: "Parkersburg was certainly contaminated with chemicals. C8, right?"

> Response from DuPont: "Ah...um...I don't know if I would characterize it that way."

> Question from Plaintiffs: "Fifty-thousand pounds annually put into the river is not a contamination?"

> Response from DuPont: [Silent, blinking][276]

As a result of this litigation, a class action settlement fund was set up to cover the costs of blood testing 70,000 residents in the Ohio River Valley. These blood tests linked DuPont's contamination of drinking water with PFAS to kidney cancer, testicular cancer, thyroid disease, ulcerative colitis, high cholesterol, and pregnancy-induced hypertension in Ohio River Valley residents.[277] In case you are thinking that the Ohio River Valley is a small region, you are mistaken. The Ohio River Valley and its water supply extend across Ohio, Illinois, Indiana, Kentucky, New York, Pennsylvania, Virginia, and West Virginia.[278] The blood testing confirmed that DuPont poisoned the drinking water and the health of Americans in at least eight US states.

Hundreds of internal company documents were also produced in this litigation with DuPont. They revealed decades of corporate cover-up regarding the hazards of PFAS.[279] Robert Bilott, the corporate defense attorney turned plaintiff's lawyer who has spent a large chunk of his career litigating over PFAS contamination had the following to say, testifying in front of Congress:

---

[275] Leach v. DuPont, "Order on Class Certification and Related Motions," Civil Action No 01-C-608, Circuit Court of Wood County, West Virginia (Jan, 15, 2003); and Robert Bilott, *Exposure* (Simon & Schuster 2019).

[276] From deposition clip included in the film, Stephanie Soechtig & Jeremy Siefert, *The Devil We Know* (2018), available at https://thedevilweknow.com/.

[277] C8 Science Panel Website Home Page (updated January 4, 2017), available at http://www.c8sciencepanel.org.

[278] Ohio River Valley Water Sanitation Commission, "Commissioners," available at http://www.orsanco.org/about-us/commissioners/ (last visited October 5, 2019) (identifying states with jurisdiction).

[279] See underlying repository of 3M and DuPont litigation documents at http://www.toxic-docs.org.

[The] companies that manufacture these chemicals have been aware of the risks for decades but failed to alert the rest of us. I know because I spent the last 20 years of my career in litigation...pulling out of their own internal files what was already there and what was already known [for decades] about the risks of these chemicals....[280]

Litigation turned over another rock; DuPont had successfully recruited the EPA to help cover up the dangers of PFAS.[281] For example, the company explicitly asked the EPA to issue a public statement that consumer products sold under the Teflon brand name are safe. The EPA was happy to do so in 2006, issuing the following statement:

The [EPA] does not believe that consumers need to stop using their cookware, clothing, or other stick-resistant, stain-resistant products.[282]

That same year, the US EPA did ask eight major US manufacturers (including DuPont) to voluntarily phase out the use of PFOA by the year 2015.[283] The EPA has done nothing, however, for the thousands of other PFAS chemicals on the market. Nor did the EPA restrict the manufacture or use of PFOA by other manufacturers. Great fan-fare has been made around the US EPA voluntary phase-out program to make it appear that the US is regulating PFAS. However, the US is not regulating PFAS at all.

Moreover, the EPA never charged DuPont or 3M for criminal miscon-duct with respect to poisoning the residents of the Ohio River Valley, the residents of Minnesota, or the entire globe. The EPA did fine DuPont $16.5 million for failing to report the hazards of its PFOA. This is approximately 0.06% of DuPont's annual revenue at the time.[284] To put things in perspective

---

[280] Testimony of Rob Bilott, Taft, Stettinus & Hollister LLP in front of US House of Representatives Committee on Oversight & Reform (116th Congress), during the "The Devil They Knew: PFAS Contamination and the Need for Corporate Accountability, Part II" Hearing (Sept. 10, 2019); Rob Bilott's litigation with DuPont is the subject of a recent movie by Mark Ruffalo, et al., *Dark Waters* (2019).

[281] Robert Bilott, *Exposure*, ch 29 (Simon & Schuster 2019).

[282] Sharon Lerner, "The Teflon Toxin: Part 3," *The Intercept* (Aug. 20, 2015) and the documen-tary Stephanie Soechtig & Jeremy Seifert, *The Devil We Know* (2018), available at https://thedevilweknow.com/.

[283] 2016 US EPA, "EPA's Non-CBI Summary Tables for 2015 Company Progress Reports (Final Progress Reports)," available at https://www.epa.gov/sites/production/files/2017-02/documents/2016_pfoa_stewardship_summary_table_0.pdf (last visited Dec 23, 2019); US EPA "Fact Sheet: 2010/2015 PFOA Stewardship Program," available at https://www.epa.gov/assessing-and-managing-chemicals-under-tsca/fact-sheet-20102015-pfoa-steward-ship-program#action (last visited Dec. 23, 2019).

[284] US EPA, "E.I. DuPont de Nemours and Company PFOA Settlements" available at https://www.epa.gov/enforcement/ei-dupont-de-nemours-and-company-pfoa-settle-ments (last visited Oct. 7, 2019); Fortune 500, "A database of Fortune list of America's

The EPA fined DuPont the equivalent of $36 for someone who makes $60,000 per year. The fine for poisoning the world is apparently not as expensive as a parking ticket.

CHAPTER 12: MORE FOREVER CHEMICALS

DuPont ultimately replaced PFOA with another PFAS chemical that is commonly known as GenX. According to the US EPA's own fact sheets, "[a] nimal studies have shown health effects [from GenX] in the kidney, blood, immune system, developing fetus, and especially in the liver following oral exposure. The data are suggestive of cancer."[285] Moreover, according to the US EPA, "GenX chemicals have been found in surface water, groundwater, finished drinking water, rainwater, and air emissions."[286] Despite these problems, the EPA issued a statement that it does not plan to regulate GenX "at this time."[287]

DuPont knows that PFAS chemicals are toxic and has decided not to stick around until the public figures this out and forces the federal government to do something about it. The company spun off a separate entity called Chemours to manufacture GenX and other PFAS chemicals, marketing Chemours as a "new startup with 200 years of experience."[288] DuPont can

[285] "Fact Sheet: Draft Toxicity Assessments for GenX Chemicals and PFBS," US EPA (November 2018), available at https://www.epa.gov/sites/production/files/2018-11/documents/factsheet_pfbs-genx-toxicity_values_11.14.2018.pdf.

[286] US EPA "Basic Information About PFOAs" (last visited November 23, 2019), available at https://www.epa.gov/pfas/basic-information-pfas#important; US EPA, "Fact Sheet: Draft Toxicity Assessments for GenX Chemicals and PFBS" US EPA (November 2018), available at https://www.epa.gov/sites/production/files/2018-11/documents/factsheet_pfbs-genx-toxicity_values_11.14.2018.pdf.

[287] "Fact Sheet: Draft Toxicity Assessments for GenX Chemicals and PFBS" US EPA (November 2018) available at https://www.epa.gov/sites/production/files/2018-11/documents/factsheet_pfbs-genx-toxicity_values_11.14.2018.pdf.

[288] Chemours, "A New Startup with 200 Years of Experience," YouTube (Jun 29, 2015), available at https://www.youtube.com/watch?v=SgpKDhLW0LU; and see Testimony of Daryl Roberts, Chief Operations and Engineering Officer, DuPont De Nemours, Inc.,

now purchase its PFAS from Chemours without assuming liability for manufacturing these chemicals. And DuPont can potentially shirk its responsibility for clean-up of PFAS contamination from the past. At least this is what DuPont argued should be the outcome during a September 2019 Congressional hearing.[289] At the same hearing, Chemours told Congress that it does not have the financial resources to assume DuPont's liability for clean-up.[290]

Chemours is nevertheless producing more PFAS and introducing them into our environment. In February 2019, independent journalists revealed that Chemours has been importing GenX toxic waste from the Netherlands to North Carolina and Texas for disposal.[291] The North Carolina DEQ wrote a letter of objection to Chemours demanding an explanation for this conduct. Chemours responded that the toxic waste was considered "safe" by federal regulators and that the US EPA had already been notified:[292]

> Please note that the spent [GenX waste] is safe under [the current federal law] and that Chemours (and DuPont before it) has notified the [EPA] of these reclamation activities on multiple occasions. Chemours manages the reclamation of the spent [waste] similar to how it manufactures [the material].

The EPA subsequently confirmed that Chemours has been "legally" dumping GenX waste in the US since 2014.[293] In light of the public backlash, however, the EPA reluctantly decided to write Chemours a "temporary"

in front of US House of Representatives Committee on Oversight & Reform (116th Congress), during the "The Devil They Knew: PFAS Contamination and the Need for Corporate Accountability, Part II" Hearing (Sept. 10, 2019).

[289] See Testimony of Daryl Roberts, Chief Operations and Engineering Officer, DuPont De Nemours, Inc., in front of US House of Representatives Committee on Oversight & Reform (116th Congress), during the "The Devil They Knew: PFAS Contamination and the Need for Corporate Accountability, Part II" Hearing (Sept. 10, 2019).

[290] See Testimony of Paul Kirsch, President of Flouroproducts, The Chemours Company, in front of US House of Representatives Committee on Oversight & Reform (116th Congress), during the "The Devil They Knew: PFAS Contamination and the Need for Corporate Accountability, Part II" Hearing (Sept. 10, 2019).

[291] Sharon Lerner, "Chemours is Using the U.S. as an Unregulated Dump for Europe's Toxic GenX Waste." *The Intercept* (February 1, 2019) & attached US EPA "Notice of Temporary Objection," available at https://www.documentcloud.org/documents/5699173-EPA-Notice-Objection-Chemours-12-19-2018.html#document/p1 (December 19, 2018).

[292] Sharon Lerner, "Chemours is Using the U.S. as an Unregulated Dump for Europe's Toxic GenX Waste." *The Intercept* (February 1, 2019) & attached letter from Chemours to North Carolina Department of Environmental Quality, dated January 24, 2018, available at https://www.documentcloud.org/documents/5699461-Chemours-Letter-to-NC-DEQ.html.

[293] Sharon Lerner, "Chemours is Using the U.S. as an Unregulated Dump for Europe's Toxic GenX Waste." *The Intercept* (February 1, 2019) & attached US EPA "Notice of Temporary Objection" (December 19, 2018), available at https://www.documentcloud.org/documents/5699173-EPA-Notice-Objection-Chemours-12-19-2018.html#document/p1.

objection letter. In its letter, the EPA asked Chemours for detailed information about the company's activities in the US. As paraphrased by me, these questions have the following flavor:[294]

- What is being shipped into the US?
- How much is being shipped?
- How toxic is this stuff?
- How much of it are you dumping?
- How are you dumping it?
- Where are you dumping it again?

One is left to wonder: How could have the EPA agreed to the dumping of GenX waste in the US as early as 2014 but not know the answer to these questions?

Across the Atlantic, DuPont's Chemours had also been dumping GenX waste in Italy until it was discovered that the waste had contaminated the drinking water supply. Italian authorities announced a national emergency and opened an investigation that has led to criminal charges.[295]

PFAS contamination has grown into a national crisis in the US because of the ubiquitous use of the chemicals in consumer products, manufacturing, military training, and firefighting.[296] The Detroit Free Press calls PFAS contamination Michigan's biggest environmental crisis in 40 years.[297] Many other states have begun testing their waterways for contamination as well, with over 1000 contaminated sites already having been identified in the US.[298] Tired of waiting for the EPA to do something about the growing PFAS problem, New Jersey has decided to set its own drinking water standards for PFOA.[299]

---

[294] Sharon Lerner, "Chemours is Using the U.S. as an Unregulated Dump for Europe's Toxic GenX Waste." *The Intercept* (February 1, 2019) & attached US EPA "Notice of Temporary Objection" (December 19, 2018), available at https://www.documentcloud.org/documents/5699173-EPA-Notice-Objection-Chemours-12-19-2018.html#document/p1.

[295] Sharon Lerner, "Chemours is Using the U.S. as an Unregulated Dump for Europe's Toxic GenX Waste." *The Intercept* (February 1, 2019).

[296] "PFAS Chemicals Must Be Regulated as a Class, Not One by One," *Environmental Working Group* (May 6, 2019).

[297] Keith Matheny, "PFAS contamination is Michigan's biggest environmental crisis in 40 years," *Detroit Free Press* (April 26, 2019).

[298] "PFAS Contamination in the United States: "Mapping the PFAS Contamination Crisis: New Data Show 1,398 Sites in 49 States," *Environmental Working Group* (last visited Dec. 23, 2019).

[299] "Affirming National Leadership Role, New Jersey Proposes Stringent Drinking Water Standards for PFOA and PFOS." State of New Jersey Department of Environmental Quality (April 1, 2019).

DuPont and 3M have yet to clean up our waterways. 3M reached a $890M settlement with the State of Minnesota to clean up water contamination from its production facilities there. While this sounds like an impressive financial commitment on the part of 3M, it is not a huge burden on its financial coffers: 3M made $32.8B in revenue in 2018 alone.[300] Moreover, the company is aggressively fighting any attempts to force it to clean up sites in other states. During the September 2019 Congressional hearing, 3M's PR representative denied that PFAS—including PFOA—were even dangerous.[301] While this position outraged an entire room full of legislators, it was the only strategy left for 3M. The company faces potential liability for PFAS contamination across the country and has run out of legal excuses.

Congressman Dan Kildee from Michigan had the following to say about the 3M, DuPont & Chemours political circus around PFAS:

> This is ridiculous...[In Oscoda, Michigan as a result of PFAS contamination] you can't hunt the animals because the groundwater has contaminated them. You can't eat the fish that you catch because they are too dangerous to consume. And we have companies who have benefited and have made millions and billions of dollars by selling these products into commerce who now want to point the finger at somebody else or say, 'well we're not going to produce these chemicals anymore—but believe me there's no science that says they're [dangerous]' ...There's plenty of science out there that demonstrates that these are harmful chemicals and dangerous for human consumption....[302]

Right now is a critical period in US politics around PFAS. The September 2019 hearing was held to help decide whether Congress should pass legislation regulating PFAS. The hearing also involved inquiry into whether Congress should designate PFAS as hazardous substances to be cleaned up under the Comprehensive Environmental Response, Compensation, and Liability Act of 1980 (aka CERCLA or Superfund).

---

[300] 3M *Annual Report* (2018).

[301] See Testimony of Denise R. Rutherford, Senior Vice President of Corporate Affairs, The 3M Company in front of US House of Representatives Committee on Oversight & Reform (116th Congress), during the "The Devil They Knew: PFAS Contamination and the Need for Corporate Accountability, Part II" Hearing (Sept. 10, 2019). 3M's Corporate Affairs organization "brings a strategic, proactive approach to advancing and protecting the 3M brand and reputation." 3M Website - Person Details: Denise R. Rutherford, available at https://investors.3m.com/governance/board-of-directors/person-details/default. aspx?ItemId=e6c0744f-44eb-4266-8e26-354045a6cbe1 (last visited Dec. 22, 2019).

[302] See Testimony by Congressman Dan Kildee in front of the US House of Representatives Committee on Oversight & Reform (116th Congress), during the "The Devil They Knew: PFAS Contamination and the Need for Corporate Accountability, Part II" Hearing (Sept. 10, 2019).

Between 3M, Chemours, and DuPont, only DuPont is on-board with the idea of cleaning up PFAS. This is because DuPont does not view itself as having any legal obligation to pay for clean-up, punting over to Chemours. This is highly convenient for DuPont, but a problem for the rest of us because Chemours is a much smaller company and does not have the resources to begin to clean up DuPont's mess.

Moreover, DuPont's motivation to support clean-up is highly profit-motivated. The company has been developing an extensive wastewater filtration portfolio.[303] According to a 2019 report, DuPont is one of the leading players in a growing market for water filtration chemicals, which is expected to have an unprecedented return on investment in the next decade.[304] Indeed, DuPont recently announced a global price increase of up to 20% on its services, exploiting the growing demand for clean-water in a polluted world.[305] And it wants to obtain the PFAS clean-up contracts that will inevitably come through as a result of our environmental disaster.

In DuPont's ideal world, Chemours, 3M, and the public will pay for clean-up, while DuPont will get paid for doing the clean-up work. Just like with Bayer-Monsanto (Chapter 10), this polluter is preparing to profit from its own pollution. On the one hand, DuPont's decision to move into the water filtration business is a positive thing. It is an actual step on DuPont's part to move into a more sustainable line of business. However, spinning off Chemours and absolving DuPont of legal responsibility for past and future use of PFAS is not helping any of us. The company owes society for clean-up, as well as the health problems it has caused and continues to cause.

Moreover, Chemours and 3M owe it to us to replace PFAS with safer chemicals. Given our history, it is unlikely that the EPA will take meaningful action to make this happen. Robert Bilott, the attorney who litigated against DuPont for over twenty years, has noted that there are feasible alternatives to PFAS that have been discussed within DuPont. The company has simply chosen not to pursue them.[306] Bilott is also skeptical about the EPA's ability

---

[303] "DuPont to Acquire Membrane Business from Evoqua Water Technologies, Corp," DuPont Press Release, available at https://www.dupont.com/news/dupont-to-acquire-membrane-business-from-evoqua-water-technologies-corp.html (Oct. 3, 2019); see also Michelle Caffrey, "DuPont to Buy Water Filtration Product Line for $110M," *Philadelphia Business Journal* (Oct. 3, 2019). DuPont waited until after the congressional hearings to announce a huge acquisition BASF's water filtration business.

[304] Press Release, "Waste Water Treatment Chemicals Market Next Big Thing," *MarketWatch* (Nov. 4, 2019).

[305] "DuPont Water Solutions Announces Global Price Increase For Ion Exchange And Reverse Osmosis Products," *Water Online* (January 15, 2019); Press Release, "World Water Development Report 2019," United Nations (March 18, 2019).

[306] Robert Bilott, *Exposure*, 283-284, 347-348 (Simon & Schuster 2019); see also "Poisoned Water & Corporate Greed: Attorney Robert Bilott on His 20-Year Battle Against

to move forward with regulation. Exasperated by the decades of inaction, he recently filed a class action lawsuit on behalf of all Americans exposed to PFAS to step up the pressure.[307]

There are strong forces at work to keep our government from regulating PFAS, and these forces keep coming out of the woodwork. The biomedical devices industry, for example, has taken a strong position that PFAS should not be regulated.[308] They argue that PFAS chemicals used for medical devices such as PTFE (aka Teflon and Gore-Tex implants) are fluoropolymers and are distinguishable from other PFAS chemicals such as PFOA.[309] Unfortunately, the medical device industry's claims of safety are contrary to a growing body of scientific evidence on this subject.[310] PFAS are simply not safe. And PFAS implants will inevitably join the long list of human implants that have been recalled for causing problems such as cancer and autoimmune illness.[311]

DuPont," *Democracy Now* (January 23, 2020).

[307] See Robert Bilott, *Exposure* 369-370 (Simon & Schuster 2019); see also "Poisoned Water & Corporate Greed: Attorney Robert Bilott on His 20-Year Battle Against DuPont," *Democracy Now* (January 23, 2020).

[308] "Medical device makers press lawmakers to drop PFAS measures" *InsideEPA.com* (Sept. 6, 2019); John Breslin, "Manufacture of life-saving devices under threat from PFAS legislation, industry says," *Legal Newswire* (Sept. 30, 2019).

[309] Barbary Henry, et al. "A Critical Review of the Application of Polymer of Low Concern and Regulatory Criteria to Fluoropolymers," *Integrated Environmental Assessment and Management* 14:3, 316-334 (January 2018); Drake Bennett, "The Company Behind Gore-Tex is Coming for your Eyeballs" *Bloomberg Businessweek* (May 9, 2019), noting "PTFE was by no means obscure: Discovered decades before at DuPont, it was commercialized in 1945 under the brand name Teflon," and discussing applications for expanded PTFE (ePTFE).

[310] See e.g., Joao Ferreira, et al., "Evaluation of Surgically Retrieved Temporomandibular Joint Alloplastic Implants - Pilot Study," *Journal of Oral Maxillofacial Surgery* 66(6) (June 2008); "Paul W. Flint, "Comparison of Soft Tissue Response in Rabbits following Laryngeal Implantation with Hydroxylapatite, Silicone Rubber, and Teflon," *Annals of Otology, Rhinology & Laryngology* 106(5) (May 1997); Arthur H. Rotstein, "TMJ Victims Blame Teflon Jaw Implants for Much Pain and Suffering : Medicine: Device, now off the market, was intended to cure an excruciating ailment. Inventor moved to Switzerland to escape lawsuits; firm is bankrupt," *Los Angeles Times* (July 3, 1994); Ryota Kikuchi, "Diffuse alveolar hemorrhage after use of a fluoropolymer-based waterproofing spray," *SpringerPlus* 4, 270 (June 2015); Namhoon Lee, "Pneumoconiosis in a polytetrafluoroethylene (PTFE) spray worker: a case report with an occupational hygiene study," *Annals of Occupational and Environmental Medicine* 30, 37 (June 2018).

[311] For a full list of implant recalls, I recommend looking at the ICIJ online database of implant recalls, safety notices, and safety alerts—there are over one hundred thousand entries. International Consortium of Investigative Journalists (ICIJ), "International Medical Devices Database," available at https://medicaldevices.icij.org/

Chapter 13: A Public Nuisance

The mysterious ingredient known as "fragrance" can be found in laundry detergent, soap, shampoo, body lotion, toilet paper, floor cleaner, body lotions, air fresheners, deodorants, and many other products. Companies entice us to buy their products, attracted to the scents they design for us. We are persuaded that our laundry, hair, homes, and vehicles need to smell like lavender, the ocean, or cherry vanilla. But nothing about the chemicals in fragrance has anything to do with these things.

"Fragrance" is a synthetic chemical cocktail and there can be several dozen to several hundred chemicals in the formulation of any given product.[312] Extensive research reports done by consumer advocacy groups have documented the toxic chemicals found in fragrance. These toxic chemicals include hormone disruptors, carcinogens, neurotoxins, petrochemicals and chemicals used in pesticides.[313] Phthalates, a class of chemicals used to make scents last longer, have widely been studied for their hormone disrupting properties and are associated with thyroid issues, diabetes, obesity, neurological problems, and reproductive disorders.[314]

---

[312] Bickers DR, et al. "The safety assessment of fragrance materials." *Regulatory Toxicology & Pharmacology* 37(2), 218–273 (April 2003).

[313] "Unpacking the Fragrance Industry: Policy Failures, the Trade Secret Myth and Public Health" Women's Voices for the Earth (updated Sept. 2018); "Not So Sexy" *Environmental Working Group and the Campaign for Safer Cosmetics* (May 12, 2010); "Perfume: An Investigation of Chemicals in 36 Eaux de Toilette and Eaux de Parfum," Greenpeace (2005); Robin E. Dodson, et al., "Endocrine Disruptors and Asthma Associated Chemicals in Consumer Products," *Environmental Health Perspectives* 120(7), 935-943 (July 2012).

[314] A.C. Gore, et al., "EDC-2: The Endocrine Society's Second Scientific Statement on Endocrine-Disrupting Chemicals," *Endocrine Review* 36(6): E1–E150 (Dec 2015); Thaddeus T. Schug, et al., "Minireview: Endocrine Disruptors: Past Lessons and Future Directions,"

Such toxic chemicals off-gas from scented products continuously. If we are not absorbing them directly through our skin, we are inhaling these chemicals into our sinuses and lungs. And a relatively small exposure to fragrance can set people off:

> I find some [scents] give me a headache [instantly].

> I can't go to church because 1 am overwhelmed by all of the perfumes, colognes and aftershaves.

> [T]his stuff needs to be banned from ALL public restrooms. The air freshener is the worst, gives me major headaches. I have gotten used to bringing my own soap/sanitizer but would be nice to not have to anymore!

> —Anonymous Victims

Unfortunately, the fragrance industry is not moved by these issues because they are making lot of money. And it is almost an entirely unregulated industry; it need not disclose its ingredients to us or even their direct clients—the personal care and cleaning product companies who market fragranced products to us.[315]

Companies like Procter & Gamble and Unilever have made great fortunes pushing fragrance onto the world, creating products that have no other purpose than to shower us with synthetic scents. There is Gain Laundry Detergent + Aroma Boost. According to one retail website, it contains "Perfume Micro Capsules that gradually break and release scent while you wear your clothes so you're always getting bursts of" synthetic fragrance.[316] Customer reviews on the Procter & Gamble website attest that the scent lasts for weeks in their clothes.[317] The chemicals last because they are persistent in the environment—they are also very difficult to break down by your body.

If that is not enough scent for you, you can add on products such as Gain Fireworks and Gain Fabric Softener to increase the concentration of synthetic chemicals in your world. The fragrance chemicals in our laundry cleaning products are so substantial that they significantly contribute

---

Molecular Endocrinology, 30(8), 833–847 (August 2016); "Endocrine Disruptors: from Scientific Evidence to Human Health Protection" 11, PE 608.866, European Union Policy Department for Citizens' Rights and Constitutional Affairs (May 2019).

[315] "Unpacking the Fragrance Industry: Policy Failures, the Trade Secret Myth and Public Health"Women's Voices for the Earth (updated Sept. 2018).

[316] "Gain + Aroma Boost Laundry Detergent," Rite Aid Website (last visited Jan. 10, 2020).

[317] "Gain + Aroma Boost Laundry Detergent " P&G Everyday Website (last visited Jan. 10, 2020).

to poor outdoor air quality. One study found that chemicals known to be hazardous air pollutants and those known to cause cancer are emitted from residential drier vents at significant quantities.[318] Laundry fragrance is such a persistent chemical cocktail that it can also be smelled on people's clothes up to 100 feet away. This becomes especially obvious if you are hiking in the woods—sometimes a laundry scent precedes people before you even see them on the trail. It is the last thing you want to be smelling, however.

Frequently the same companies who peddle fragranced laundry detergent also peddle chemical air fresheners. Procter & Gamble's Febreze air freshener, for example, comes in liquid spray, wax melts, candles, and other concoctions you can easily inhale. "Crank up the *fresh-tensity* of your home's air and clean away stinks," advises Procter & Gamble's marketing team.[319] These products do not actually do any cleaning of the air, however. Nor do they eliminate the source of the stink in your home. Rather, they cover up the stench of your rotting garbage, cigarette smoke, or dog puke with an intense blast of persistent and problematic chemicals.[320]

Fragranced products are also reported to contain nerve-damaging chemicals that deaden your sense of smell.[321] So you no longer smell the dog puke—although it is still there. And you cannot smell the fragrance very well anymore. As a result, you will buy even more Febreze or similar products to have a "fresh scent" experience. Rather than poisoning yourself, it is much more effective to get rid of smells by changing the trash often, use baking soda to absorb odors, vinegar to break down body fluids, wet mopping with water, and opening the window.

There is a lot of greenwashing going on around fragrance. Many personal care and household products claim the benefits of natural ingredients but still contain synthetic fragrance. Companies sometimes obscure their use

---

[318] Anne Steinemann, et al., "Chemical emissions from residential dryer vents during use of fragranced laundry products," *Air Quality, Atmosphere & Health* 6, 151-156 (March 2013).

[319] "Febreze Unstoppables Wax Melt Fresh," Procter & Gamble Febreze Website (last visited Jan. 10, 2020).

[320] Bickers DR, et al.. "The safety assessment of fragrance materials." *Regulatory Toxicology & Pharmacology* 37(2), 218-273 (April 2003). "Unpacking the Fragrance Industry: Policy Failures, the Trade Secret Myth and Public Health"Women's Voices for the Earth (updated Sept. 2018); "Not So Sexy" *Environmental Working Group and the Campaign for Safer Cosmetics* (May 12, 2010); "Perfume: An Investigation of Chemicals in 36 Eaux de Toilette and Eaux de Parfum," Greenpeace (2005); Robin E. Dodson, et al., "Endocrine Disruptors and Asthma Associated Chemicals in Consumer Products," *Environmental Health Perspectives* 120(7), 935-943 (July 2012).

[321] "Avoid Harsh Chemicals in Commercial Air Fresheners with Homemade Alternatives," *Scientific American* (September 9, 2012); Armida Stickney, "Petition for a Consumer Product Safety Rule Governing Febreze," addressed to Marietta S. Robinson, Commissioner, U.S. Consumer Product Safety Commission (Nov. 22, 2013), available at https://maskedca-naries.files.wordpress.com/2014/01/cpsc-petition-re-febreze-rev-102713.pdf.

of synthetic chemicals by packaging their product in eco-friendly materials, such as recycled cardboard or glass. We pick up on the visual cues and assume the product is natural because the packaging looks natural. Unless we look at the ingredient list, we will never know.

There is also a lot of misinformation around the safety of essential oils, which are marketed as an alternative to synthetic chemicals. People assume that products scented with essential oils are safe. However, problematic chemicals are still sometimes mixed in with essential oils to keep the scents long-lasting. And essential oils themselves have been proven to off-gas chemicals that are toxic to us.[322] These chemicals may be relatively benign when we smell them off of a plant or flower, but become problematic when used in the concentrations we get by producing essential oils. It takes thousands of flowers to produce a bottle of essential oil. That type of concentration is quite irritating to our nasal passages, lungs, nervous system, and our cells. Indeed people with serious chemical sensitivity and/or chronic fatigue often complain that even essential oils can trigger chronic fatigue and other symptoms.

Just like with cigarettes, a variety of fragranced products are marketed aggressively to teenagers and children. Kids are known to be much more easily influenced by advertising. And companies wish to create habits that will make them life-long customers. The poster-child example for this is Axe Body Spray. The California Air Resources Board fined Unilever, the maker of Axe Body spray, $1.3 million for violating air quality standards with the product.[323] It has an intense odor that most adult men would gladly avoid. Teenage boys, however, are led to believe through advertising that the stench will attract women. It is unconfirmed to do so; but it has sent numerous teens to the emergency room with acute allergic reactions.[324]

Fragrance has become a major pollutant, but a liberally used substance in our society.[325] Airlines use fragrance to cover up the diesel exhaust inadvertently seeping into the main cabin and to mask the smell of their uncleaned toilets. Taxi, Uber and Lyft drivers frequently have air fresheners in their cars to mask the smell of last night's vomit. Hotels wash their sheets and

---

[322] Neda Nematollahi, et al., "Volatile Chemical Emissions from Essential Oils," *Air Quality, Atmosphere & Health* 11, 949-954 (August 2018).

[323] "Unilever/Conopco Settlement," California Air Resources Board (January 2010).

[324] "Axe Body Spray Shuts Down School, Sends 8 Students to Hospital," *HuffPost NYC* (July 28, 2014); Stefani Forster, "Axe Body Spray Allergy: Teen Hospitalized After Allergic Reaction," *HuffPost Canada* (March 3, 2018). See also Joelle Goldstein, "Florida School Bus Evacuated Over 'Hazardous Materials'—But It Turns Out to Be Axe Body Spray," *People* (Dec. 12, 2019).

[325] Anne Steinemann, et al., "Fragranced consumer products: chemicals emitted, ingredients unlisted," *Environ Impact Assessment Review* 31(3), 328–333 (April 2011).

clean their rooms with fragranced products to signal to you that the room has been serviced. Grocery and drug stores carry such a large concentration of fragranced products that they unwittingly circulate fragrance throughout their indoor air supply.

Hotels and clothing stores have started to pump out fragrance through the central circulation system, blasting people with their "signature scent." This is a marketing ploy intended to have you walk into the building and feel like you have entered somewhere special, exciting, and maybe even sexy. Many people find these chemicals and fragrances so overwhelming, however, that they feel chased away:

> [F]ragrance in the vents is nothing new at Target. Especially in winter, you can just stand outside the front doors, and when they open, and you get a blast of warm air, the Target signature scent is so strong. I haven't been able to go inside there in years.
>
> —Anonymous

> Every Target around me is really bad with the signature scent blowing through the HVAC. I smell it 100 feet away from the door out in the parking lot. Even if I wore a mask the smell sticks to my hair and clothes. And it soaks into all the merchandise I bring home, including grocery items (so gross.)
>
> —Anonymous

Target and other stores may want you to feel like you are somewhere special, exciting, and even sexy. But the reality is that you are still just shopping at a store. And getting exposed to toxic chemicals in the process.

> I believe this is a civil right violation. I should not be covered in sticky fragrant crap while trying to shop at Costco...or a Target, but I am. Nobody has a right to spray fragrance on me against my will... .I'm not sure Costco has started putting fragrances in the AC system, but they are selling "Unstoppable" scent enhancers and other sticky fragrant laundry products... I get home and...my clothes are giving off fragrance smell and now I need to wash my clothes...take a shower & wash my hair....
>
> —Anonymous

If you work in one of these establishments, you have no choice but to be exposed to fragranced products day in and day out. If you work as a flight attendant, you will be forced to spend countless hours on planes pumped full of fragrance. The same issue arises for office workers who might be exposed to fragrance from their colleagues' perfume or laundry detergent, or whenever they encounter the air freshener in the office bathroom. All of these individuals have no choice but to be engulfed in illness-inducing scented products in order to keep their jobs. This is an unfair choice that no one should have to make.

> I don't like to smell any sh*** fragrance. The grocery stor[e] carts always have some sh*** fragrance from somebody, I have to wipe the handle of the cart, otherwise somebody [else's] fragrance contaminates my hands, purse, etc. My previous [boss] used to give me paperwork to file, and the whole paperwork was contaminated by her perfume, [which ended up all over] my hands....

> —Anonymous

It is no wonder that a number of recent articles have called fragrance the new second hand smoke.[326] People once smoked in restaurants, hotels, taxis, and even on airplanes. Smoking became a major source of poor indoor air quality. Employees of bars and restaurants were especially hit hard, as they had no choice but to endure the smoke to their health detriment. We ultimately banned smoking from virtually all buildings and we need to ultimately ban fragrance as well. While this may sound like an ambitious proposition, it actually aligns with popular opinion. A national survey reported the following regarding people's attitudes toward fragrance:[327]

- Over half of the American population would prefer that their workplaces, hotels, airplanes, and health care facilities be fragrance free.
- 34% reported health problems from fragrance (including headaches, respiratory problems, skin problems, digestive issues, among others).
- 20.2% reported leaving a business as soon as possible upon smelling air fresheners or a fragranced product.
- 17.5% were reluctant or unable to use public toilets because of the scented bathroom products.
- 14.1% were reluctant or unable to use soap in a public bathroom for this reason.
- 15.1% reported getting sick at work, losing workdays or quitting their job because of the use of fragrance there.

---

[326] See e.g., Elisabeth Leamy, "Bothered by fragrances? This story will be a breath of fresh air." *The Washington Post - Home and Garden Section* (March 20, 2018).

[327] Steinemann, Anne, "Fragranced consumer products: exposures and effects from emissions," *Air Quality Atmosphere & Health* 9(8), 861–866 (Oct 2016).

And these people are very frustrated by their resulting serious health challenges:

> Fragrance can cause serious health issues-migraines, brain fog, extreme fatigue- these are just a few of the symptoms I feel when fragrances overwhelm my system... it's not a good feeling. Please consider others before you spray yourself in perfume or fragrances....
>
> —Anonymous Victim

> I told my [supervisor I am highly sensitive to fragrance] and then he comes to work smelling of cologne. I've had to leave work and go to ER several times. It got to the point where I didn't want to fight anymore. I decided to put my health first and go on short term disability.
>
> —Anonymous Victim

We have trouble wrapping our heads around the fact that other people could become ill as a result of our fragranced laundry detergent or our perfume. We believe that if people do not immediately get sick from a substance, then it must be harmless. As a result, people with fragrance sensitivity are frequently socially ostracized and—oddly enough—verbally abused. For some reason, many individuals can get very angry when you tell them you have a sensitivity to the scented products they are using.

The end result is that people with fragrance sensitivity are often forced to escape the stench in silence. Very few are fortunate enough to take advantage of disability laws to protect themselves. Most will simply quit their job or stop attending social and civic events in order to avoid exposure. They disappear into the background and become silenced by our toxic chemical world. Having the human canaries disappear from public life makes it even easier for industry to continue to push fragrance and other toxic chemicals our way, business as usual.

The outbreak of COVID-19 has made the situation worse. In the clutches of fear, retailers and individuals alike are reaching for fragranced chemical cleaning products and disinfectants as a way to assure themselves that they are managing disease risk. These chemical cocktails can disrupt human hormones, damage the metabolism, and undermine the immune system. While many people may not directly feel or understand the impacts of this chemical war on COVID-19, the chemically sensitive are deeply suffering. For some, it has become impossible to make a trip to the grocery store without falling ill.

Chapter 14: The Cheapest Commodity

I cannot write about fragrance in this book without discussing plastics. Indeed, plastics and fragrance have a lot in common. Plastics contain some of the same toxic chemicals used in fragrance. And a whole host of toxic chemicals are found in plastics. These chemicals are used around our food, buildings, apparel, electronics, children's toys and even hospital IV bags.[328] Some of the worst offenders are the chemicals used in vinyl; Styrofoam cups; polycarbonate bottles; and food packaging.[329] The most popular synthetic fibers used in our clothes—Nylon, Polyester, Acrylic, and Lycra/Spandex— are also technically plastic and equally troublesome.[330]

---

[328] Mark S. Rossi & Ann Blake, "The Plastics Scorecard (Version 1.0)," Clean Production Action (July 1, 2014); Joanna Malaczynski, "Chemicals of Concern in Consumer Products—Where Are They?," DESi Potential (March 20, 2020); see also the State of Washington Department of Ecology Children's Safe Product Act, "Manufacturer Reporting," available at https://ecology.wa.gov/Waste-Toxics/Reducing-toxic-chemicals/Childrens-Safe-Products-Act (last visited April 29, 2020 (self-reported data by industry of priority chemicals present in products marketed toward children, such as shoes, clothes, furniture, toys, etc.); James T Brophy, et al., "Breast cancer risk in relation to occupations with exposure to carcinogens and endocrine disruptors: a Canadian case–control study," *Environmental Health* 11, 87 (Nov. 2012); "Hidden Plastics Including Chewing Gum Made of Plastic," Friends of the Earth (last visited March 2, 2020). Moreover, a landscape architect working on new development in the Central Valley of California showed me multiple pictures of styrofoam blocks used under soil to create elevated berms in an otherwise flat suburban landscape.

[329] Mark S. Rossi & Ann Blake, "The Plastics Scorecard (Version 1.0)," Clean Production Action (July 1, 2014) (identifying most and least problematic types of plastics).

[330] Z Singh and S Bhalla, "Toxicity of Synthetic Fibres & Health," *Advanced Research in Textile Engineering*, 2(1), 1012-1016 (January 2017).

One of the most problematic classes of chemicals found in plastics is plasticizers.[331] They are hormone disruptors, linked to thyroid issues, autism, diabetes, obesity, neurodevelopment disorders, and reproductive/gender disorders.[332] Plasticizers are added to make soft plastics flexible and prevent hard plastics from becoming brittle.[333] And, as discussed in Chapter 13, they are in also frequently used in fragranced products.

Like fragrance, plastics are fabricated primary out of petrochemicals.[334] The petrochemicals used in plastics and fragrance are waste by-products of petroleum refining and are pushed upon us in consumer products so that industry does not have to find a way to dispose of them.[335] Rather, they can make money off of them. For this reason, petroleum-based plastics and fragrance have been pushed onto the market rather indiscriminately and at low cost.

---

[331] "Chemical Economics Handbook: Plasticizers" *IHS Markit* (May 2018); "Plasticizers and Solvents," Daihachi Chemical Co. Website (last visited Nov. 1, 2019); C. Zang, et al., "Most Plastic Products Release Estrogenic Chemicals: A potential Health Problem That Can Be Solved," *Environmental Health Perspectives*, 119(7), 989–996 (July 2011). "Polymer Additives & Plastizers - Library Listing 1,799 spectra," ThermoFisher Scientific (2007), available at http://www.thermo.com.cn/Resources/200802/productPDF_2099.pdf; "Endocrine disrupting properties to be added for four phthalates in the Authorisation List," European Chemicals Agency, ECHA/NR/19/26 (July 10, 2019); C.S. Giam & M.K. Wong, "Plasticizers in Food," *Journal of Food Protection* 50(9), 769-782 (Sept. 1987); "Plastics & Health" Center for International Environmental Law (February 2019).

[332] A.C. Gore, et al., "EDC-2: The Endocrine Society's Second Scientific Statement on Endocrine-Disrupting Chemicals," *Endocrine Review* 36(6): E1–E150 (Dec 2015); "Endocrine Disruptors: from Scientific Evidence to Human Health Protection," PE 608.866, European Union Policy Department for Citizens' Rights and Constitutional Affairs (May 2019); Theo Colborn, et al., *Our Stolen Future* (Penguin Group 1996); see also "International Cooperation: Persistent Organic Pollutants: A Global Issue, A Global Response," US EPA, available at https://www.epa.gov/international-cooperation/persistent-organic-pollutants-global-issue-global-response (last visited July 17, 2019); Amir Midovnik, "Endocrine disruptors and childhood social impairment," *NeuroToxicology* 32:2, 261-267 (March 2011); Youssef Oulhote, "Gestational Exposures to Phthalates and Folic Acid, and Autistic Traits in Canadian Children," Environmental Health Perspectives 128(2), 027004-1 – 027004-12 (February 2020).

[333] "Plasticizers in Polymers: They make your new car smell...like a new car." Polymer Solutions, Inc. (Nov. 11, 2015), available at http://www.polymersolutions.com/blog/plasticizers-in-polymers/.

[334] "Global Chemicals Outlook II," 30-34, 44, ISBN No: 978-92-807-3745-5, United Nations Environment Programme (2019); Mark S. Rossi & Ann Blake, "The Plastics Scorecard (Version 1.0)," Clean Production Action (July 1, 2014); Anne Marie Helmenstine, "Examples of Organic Chemistry in Everyday Life," *ThoughtCo* (July 3, 2019). See also chapters 21-22 and further discussion in this chapter about the connection between fragrance and plastics.

[335] Indeed, less than 50% of a given barrel of crude oil becomes fuel. The remaining materials must be diverted to other uses. "How many gallons of gasoline and diesel fuel are made from one barrel of oil?," U.S. Energy Information Administration - Frequently Asked Questions (last updated May 23, 2018); see also "How much oil is used to make plastic?," US Energy Information Administration - Frequently Asked Questions (updated June 4, 2019).

Indeed, the petroleum industry has a huge stake in the chemical industry.[336] Over 90% of all organic chemistry products—such as conventional plastics, synthetic fibers, cleaning products, fragrances, etc.—are derived from just seven (mostly toxic) petrochemicals.[337] Moreover, a significant amount of the largest chemical companies have been bought up by fossil fuel producers.[338] These companies have made huge investments in recent years to expand their capacity to process natural gas and conduct fracking operations in the US for the purpose of manufacturing plastics in particular.[339] And while the popularity of oil production may be going down due to concerns over climate change, the demand for petrochemicals for synthetic chemistry is projected to increase.[340] As demand for dirty energy stagnates, fossil fuel producers turn to plastics (and possibly fragrance) as a growing market for their revenue.[341] Given current trends, estimates predict plastics production will reach 2 billion metric tons by 2050, growing exponentially from 350 million metric tons today.[342]

You may have noticed that almost everything we consume is packaged in plastic: from our food to our toothpaste. Plastic packaging—which is for the most part is meant to be disposed of after a single use—accounts for nearly 40% of all plastics production.[343] And our use of plastics is growing. Our couches are upholstered in plastic and are adorned with fake plastic pillows. Most of our clothing is made of synthetic plastic fibers—not just our jackets and yoga pants, but also our shirts, sweaters and even underwear. Plastics are also foundational for most of our electronics—including our laptops, cell phones, and insulation for the myriad of wires in most electronic products.

---

[336] "Global Chemicals Outlook II," 36, ISBN No: 978-92-807-3745-5, United Nations Environment Programme (2019).

[337] "Global Chemicals Outlook II," 44, ISBN No: 978-92-807-3745-5, United Nations Environment Programme (2019); see also Royal Society of Chemistry, "Organic Chemistry Contributing to Flavours and Fragrances," (last visited March 2, 2020).

[338] "Global Chemicals Outlook II," 33, 36-37, ISBN No: 978-92-807-3745-5, United Nations Environment Programme (2019).

[339] "The Fracking Endgame: Locked into Plastics, Pollution and Climate Chaos," Food and Water Watch (June 2019).

[340] "Global Chemicals Outlook II," 34, ISBN No: 978-92-807-3745-5, United Nations Environment Programme (2019).

[341] Marc Yaggi & Gabrielle Segal, "The Fossil Fuel Industry's Plot to Stay Relevant is Made of Plastic," *Waterkeeper* (April. 22, 2018).

[342] "Global Chemicals Outlook II," 57, ISBN No: 978-92-807-3745-5, United Nations Environment Programme (2019).

[343] "Global Chemicals Outlook II," 59, ISBN No: 978-92-807-3745-5, United Nations Environment Programme (2019).

We can reduce our exposure to plasticizers and other harmful chemicals by eliminating plastics in our everyday lives. This is becoming harder to do, however, because plastic waste is such a persistent pollutant in our environment. Plastics do not break down; rather, they only become finer and finer particles known as micro-plastics. When plastics become micro-plastics, we inhale them as dust in our homes and swallow them in our food.

Plastic dust comes off of our clothing, furniture, blankets, upholstery, and carpets. It leaches from our washing machines and can enter the atmosphere through our clothes dryers. Micro-plastic particles also contaminate our beer, honey, sushi, and many other foods. Contamination comes from disposable packaging and most of our plastic products. We ingest micro-plastics from food—fish, for example, contain micro-plastics they have swallowed in the ocean. Micro-plastics are such a pervasive problem that some scientists estimate we are ingesting a credit-card's worth of plastic every week in the form of fine particles we cannot even see.[344] These invisible micro-plastic particles are able to penetrate even the human placenta.[345]

Nevertheless, there is plenty of plastic waste we can see. It is estimated that up to 12.7 million metric tons of plastic entered the ocean annually in 2010; the figure is estimated to reach up to 250 million metric tons annually by 2025.[346] Because industry is not responsible for getting rid of plastic waste once it has reached the consumer's hands, it has largely washed its hands of our mounting plastics pollution problem.[347]

At our current trajectory, scientists estimate that there will be more plastics in the ocean than fish by the year 2050.[348] Sadly enough, scientific reports were coming back as early as the 1970s that plastics were polluting our oceans.[349] Researchers were documenting that these substances were

---

[344] "Plastic ingestion by people could be equating to a credit card a week," University of Newcastle Australia (June 12, 2019), available at University of Newcastle Australia.

[345] Antonio Ragusa, et al., "Plasticenta: First Evidence of Microplastics in Human Placenta," 146 *Environment International* 106274 (January 2021).

[346] "Global Chemicals Outlook II," 102, ISBN No: 978-92-807-3745-5, United Nations Environment Programme (2019).

[347] Contrary to popular opinion, the vast majority of plastics are never recycled. Alexander H. Tullo, "Plastic has a problem; is chemical recycling the solution?," *Chemical & Engineering News* 96(39) (Oct. 2019). There have been some recent informal commitments by some manufacturers to reduce plastic waste from disposable containers. See Joanna Malaczynski, "2 Companies Capitalizing on the Market Shift Away from Disposable Plastic Packaging," *Sustainable Brands* (August 14, 2018).

[348] Sarah Kaplan, "By 2050, there will be more plastic than fish in the world's oceans, study says," *The Washington Post* (January 20, 2016).

[349] See e.g., Anthony Cundell, "Plastics in the Marine Environment," *Environmental Conservation* 1(1), 63-68 (Spring 1974) and Edward Carpenter & K.L. Smith, "Plastics on the Sargasso Sea Surface," *Science* 175, 1240-1241 (March 1972).

weatherizing into micro plastic particles and hurting marine life. But not much has changed.

Alternatives to petroleum-based plastics do exist. There are traditional materials such as glass, metal, wood, and many others available to us as consumers. And we also have manufactured safe, compostable bioplastics out of natural materials such as algae, hemp, banana peels, and even mushrooms. For reasons that have much more to do with the status quo rather than technological ability (Chapter 24), bioplastics have a marginal share of the market, however.

PART II: OUR CHRONIC ILLNESS EPIDEMIC

Chapter 15: The Chronic Illness Epidemic

> A living cell, like a flame, burns fuel to produce the energy on which life depends. The analogy is more poetic than precise, for the cell accomplishes its 'burning' with only the moderate heat of the body's normal temperature. Yet all these billions of gently burning little fires spark the energy of life. Should they cease to burn, "no heart could beat, no plant could grow upward defying gravity, no amoeba could swim, no sensation could speed along a nerve, no thought could flash in the human brain."....The ultimate work of energy production is accomplished not in any specialized organ but in every cell of the body.
>
> —Rachel Carson[350]

The flames that sustain life in every cell of our body referred to by Rachel Carson are collectively known as our metabolism. The energy produced by our metabolism relies upon the foods we eat, the water we drink, the sunshine we are exposed to and the air we breathe. Our metabolism converts energy into fuel for our organs, cells, and even DNA. This happens within a sequence of endless chemical reactions inside of us.

Our metabolism is regulated by our mitochondria, which are found in the DNA of our cells and which have recently been discovered to circulate in our bloodstream as well.[351] We know that mitochondria "contain 1500 proteins...

---

[350] Rachel Carson, *Silent Spring*, 200-201 (Riverside Press 1962), quoting Eugene Rabinowitch.

[351] Rachel Carson, *Silent Spring*, 202 (Riverside Press 1962); see also Al Amir Dache, et al., "Blood contains circulating cell-free respiratory competent mitochondria," *The FASEB Journal* 34(3), 3616-3630 (January 19, 2020).

and catalyze over 500 different chemical reactions" alone.[352] Excessive exposure to toxic chemicals, just like excessive exposure to radiation, damages our mitochondria and our ability to produce the energy we need to carry out the functions of our body.[353]

A single deadly chemical exposure can damage a person's mitochondria so badly that they may die, become paralyzed, or develop cancer. Often times the damage does not take full effect until weeks or months after chemical exposure, when the dysfunction in our mitochondria fully take toll throughout the internal organs, tissues, and cells. Systems slowly break down, until the body is no longer self-sustaining.

Chronic exposure to much smaller amounts of toxic chemicals is equally bad.[354] It ultimately leads to a condition known as "oxidative stress,"[355] where the cells of the body have a hard time maintaining their life-supportive environment.[356] Oxidative stress is so problematic that it actually causes visible lesions to our DNA.[357] It also disrupts our mitochondria and our life-giving metabolism.[358]

Our world has become a test laboratory for metabolic damage and oxidative stress as the use of toxic chemicals has proliferated. Environmental pollution now contaminates all of our waterways, air, soil, food, homes, workplaces, and the products we use every day. Toxic chemicals even get passed down from mother to child before birth.[359] We know that they are consistently found in our blood streams, urine, fat tissue, internal organs,

---

[352] Robert K. Naviaux, "Metabolic Features of Cell Danger Response," *Mitochondrion* 16, 7-17 (2014).

[353] Rachel Carson, *Silent Spring*, 203 (Riverside Press 1962).

[354] Rachel Carson, *Silent Spring*, 188-189 (Riverside Press 1962). As Rachel Carson explained, a "change at one point, in one molecule [of our body] even, may reverberate throughout the entire system to initiate changes in seemingly unrelated organs and tissues..."

[355] Toshikazu Yoshikawa & Yuji Naito, "What is Oxidative Stress?" *Journal of the Japan Medical Association* 45(7) 271-276 (July 2002).

[356] James M. Samet & Phillip A. Wages, "Oxidative Stress from Environmental Exposures," *Current Opinion in Toxicology* 7, 60-66 (Feb. 2018); Alexandros G. Asimakopoulos, et al., "Urinary biomarkers of exposure to 57 xenobiotics and its association with oxidative stress in a population in Jeddah, Saudi Arabia," *Environmental Research* 150, 573-581 (Oct. 2016).

[357] Bennett Van Houten, et al., "DNA repair after oxidative stress: Current challenges," *Current Opinion in Toxicology* 7, 9-16 (Feb. 2018).

[358] Helmut Sies, "On the history of oxidative stress: Concept and some aspects of current development," *Current Opinion in Toxicology* 7, 122-126 (Feb. 2018) (our life-giving metabolic processes are driven by our mitochondria and are called redox-signaling; this signaling is disrupted by xenobiotics); see also Yvonne Collins, et al., "Mitochondrial redox signalling at a glance," *Journal of Cell Science* 125, 801-806 (2012).

[359] Kirsi Vahakangas, et al., "Chapter 18 - Biomarkers of Toxicity in Human Placenta," in Ramesh C. Gupta (ed.), *Biomarkers in Toxicity, 2nd Edition* (Elsevier Press 2019); and Dana B. Barr, et al., "Concentrations of xenobiotic chemicals in the maternal-fetal unit," *Reproductive Toxicology* 23(3), 260-266 (April-May 2007).

and even breast milk.[360] Because the most troublesome of these substances do not fully break down in our systems, they also build up in our bodies.[361] And our widespread contamination with such pollutants extends to every single living being around the world, no matter how remote their location.

Under the influence of toxic chemicals, 45% of the adult US population now has at least one chronic illness.[362] And our poor health statistics are on the rise. Scientists have eliminated inactivity, diet, and other similar lifestyle factors as the cause of our rising health problems.[363] Rather, the common denominator for the aggravated rates of chronic disease is toxic chemical exposure and other forms of environmental pollution. Research has shown that that our exposure to toxic chemicals is correlated with high cholesterol, diabetes, depression, digestive disorders, liver disease, heart disease, obesity and many other health issues we commonly believe are caused by our genes or lifestyle choices.[364]

---

[360] "Fourth National Report on Human Exposure to Environmental Chemicals" US Centers for Disease Control (2019) and previous reports dating back to "National Report on Human Exposure to Environmental Chemicals" US Centers for Disease Control (2001); see also European Union, "Endocrine Disruptors: from Scientific Evidence to Human Health Protection," a study commissioned by the PETI Committee of the European Parliament PE 608.866 (March 2019).

[361] Arnaud Tonnelier, et al., "Screening of chemicals for human bioaccumulative potential with a physiologically based toxicokinetic model," *Archives of Toxicology* 86(3), 393-403 (March 2012).

[362] Wullianallur Raghupathi and Viju Raghupathi, "An Empirical Study of Chronic Diseases in the United States: A Visual Analytics Approach to Public Health," *International Journal of Environmental Research and Public Health* 15(3), 431 (March 1, 2018); individual health statistics for the US are also available from the US Centers for Disease Control, National Center for Health Statistics, available at https://www.cdc.gov/nchs/.

[363] See e.g., Swanson, N.L., Leu, A., Abrahamson, J. & Wallet, B. "Genetically engineered crops, glyphosate and the deterioration of health in the United States of America," *Journal of Organic Systems* 9 (2014) 6–37 (citing various research studies); Carla Lubrano, et al., "Obesity and Metabolic Comorbidities: Environmental Diseases?," *Oxidative Medicine and Cellular Longevity* 2013 (Feb. 5, 2013); US Centers for Disease Control, "Transcript for the 9th CDC Myalgic Encephalomyelitis/Chronic Fatigue Syndrome Stakeholder Engagement and Communication (ME/CFS-SEC)" CDC (May 25, 2017) (testimony of Robert Naviaux).

[364] Dr. Robert Naviaux, "Metabolic Features of Myalgic Encephalomyelitis/Chronic Fatigue Syndrome," presentation found in Transcript for the 9th CDC Myalgic Encephalomyelitis/Chronic Fatigue Syndrome Stakeholder Engagement and Communication (ME/CFS-SEC) (May 25, 2017), available at https://www.cdc.gov/me-cfs/pdfs/transcript-naviaux-05272017.pdf. Carla Lubrano, et al., "Obesity and Metabolic Comorbidities: Environmental Diseases?," *Oxidative Medicine and Cellular Longevity* 2013 (Feb. 5, 2013); Swanson, N.L., Leu, A., Abrahamson, J. & Wallet, B. "Genetically engineered crops, glyphosate and the deterioration of health in the United States of America," *Journal of Organic Systems* 9(2), 6–37 (January 2014); Anthony Samsel and Stephanie Seneff, "Glyphosate, pathways to modern diseases II: Celiac sprue and gluten intolerance," *Interdisciplinary Toxicology* 6(4), 159–184 (Dec. 2013).

Cancer is one of the most common chronic illnesses. Liver cancer, thyroid cancer, pancreatic cancer, breast cancer and prostate cancer—all cancers of organs most impacted by toxic chemical exposure—have grown astronomically in industrialized countries.[365] Cancer is no longer a problem for the middle-aged and older. The incidence of breast cancer in women under 40 has nearly doubled in the US in less than four decades.[366] And childhood cancer (cancer in those under the age of 20) has increased 34% since 1975.[367]

Alzheimer's disease and other forms of dementia are also a growing epidemic. Contrary to popular belief, this is not explained away by an aging population.[368] The disease is hitting the young as well as the old. Blue Cross Blue Shield recently revised its diagnosis rates of dementia-related disease in younger adults (ages 30-64) from 4.2 to 12.6 per 10,000 adults between 2013 and 2017; this is a 200% increase.[369] Alzheimer's disease is known to result from metabolic damage, insulin resistance and brain inflammation, all of which also occur in response to toxic chemical exposure.[370]

Kidney disease has also grown sharply and now affects over 30 million Americans.[371] Our kidneys eliminate waste and excess fluids from the bodies. When our kidneys become chronically overwhelmed, we will develop kidney disease. About 15% of the US population suffers from kidney disease as of 2018. Kidney disease is frequently the outcome of diabetes—another

---

[365] See e.g., Marta Benedetti et al, "Incidence of Breast, Prostate, Testicular, and Thyroid Cancer in Italian Contaminated Sites with Presence of Substances with Endocrine Disrupting Properties," *International Journal of Environmental Research and Public Health* 14(4), 355 (March 2017); Miranda M. Fidler et al, "A global view on cancer incidence and national levels of the human development index," *International Journal of Cancer* 139, 2436-2446 (Aug. 2016).

[366] Rebecca H. Johnson, et al., "Incidence of Breast Cancer With Distant Involvement Among Women in the United States, 1976 to 2009," *JAMA* 309(8), 800-805 (February 2013).

[367] "Childhood Cancer: Cross-Sector Strategies for Prevention," at 3, Childhood Cancer Prevention Initiative (2020).

[368] Liara Rizzi, et al., "Global Epidemiology of Dementia: Alzheimer's and Vascular Types," *Biomed Research International* 2014, 908915 (June 2014).

[369] Blue Cross Blue Shield, "Early-Onset Dementia and Alzheimers Rates Grow in Young American Adults" (February 2020), available at https://www.bcbs.com/the-health-of-america/reports/early-onset-dementia-alzheimers-disease-affecting-younger-american-adults.

[370] See e.g., Suzanne M. de la Monte & Ming Tong, "Brain metabolic dysfunction at the core of Alzheimer's disease," *Biochemical Pharmacology* 88(4), 548-559 (April 2014). See also the next chapters in this book, discussing metabolic disfunction, inflammation, and diabetes.

[371] Yan Xie, el al, "Analysis of the Global Burden of Disease study highlights the global, regional, and national trends of chronic kidney disease epidemiology from 1990 to 2016," *Kidney International* 93(3), 567-581 (Sept. 2018); Robert Holly, "Trump Administration Looking to Cut Dialysis Costs with Home Focus," *Home Health Care News* (March 4, 2019).

disease known to have increased steadily as a result of toxic chemicals (see Chapter 16).

Over 720,000 of Americans suffering from kidney disease have progressed to kidney failure; these are individuals wholly dependent upon kidney dialysis for survival. That means they must physically hook up to a machine to eliminate the waste and excess fluids from their bodies three times a week.[372] Forecasts expect that 1.26 million Americans will have kidney failure by 2030. Because of this growing problem, the federal government announced in July 2019 a law that will enable widespread kidney dialysis to be completed at home. A new market is burgeoning around our health woes—and the US kidney dialysis industry alone has grown to a $24 billion industry.[373]

Intestinal disorders such as Crohn's disease and colitis have increased abruptly over the last decades as well. The number of people who have been hospitalized as a result of digestive illnesses has grown sharply.[374] Many of these diseases were so uncommon in the past that they historically have had no specific disease category of their own—they were simply identified as "other digestive disorders" in government studies.[375] Children are some of the hardest hit by the new wave of intestinal disorders. In Canada, for example, the rate of inflammatory bowel disease in children under 16 years of age increased by almost 60 percent in just one decade from 1999 to 2010.[376]

The growing epidemic of digestive ailments has mostly slipped under the radar of the conventional medical community. Receiving no answer from their primary care physician, many Americans show up at the offices of acupuncturists, naturopaths, herbalists, and homeopaths complaining of problems such as intestinal bloating, irritable bowel, the inability to absorb

---

[372] Caroline Copley & Caroline Humer, "U.S. seeks to cut dialysis costs with more home care versus clinics," *Reuters* (March 3, 2019).

[373] Robert Holly, "Trump Administration Looking to Cut Dialysis Costs with Home Focus," *Home Health Care News* (March 4, 2019) (citing 2018 figures).

[374] Siew C. Ng., et al., "Worldwide incidence and prevalence of inflammatory bowel disease in the 21st century: a systematic review of population-based studies," *The Lancet* 390(10114), 2769-2778 (December 2017 – January 2018).

[375] James E. Everheart (ed), "Burden of Digestive Diseases in the United States Report," National Institute of Diabetes and Digestive and Kidney Diseases" National Institute of Digestive and Kidney Diseases (2008).

[376] Eric I Benchimol, "Trends in Epidemiology of Pediatric Inflammatory Bowel Disease in Canada: Distributed Network Analysis of Multiple Population-Based Provincial Health Administrative Databases." *The American Journal of Gastroenterology*, 112(7),1120-1134 (July 2017).

nutrients from their food, etc.[377] Trying to stem the tide, we take probiotics and digestive enzymes at rates never before seen in history.[378]

Toxic chemicals are also known to cause immune suppression, making us more vulnerable to infection and less responsive to vaccines.[379] And immune suppression from toxic chemical exposure can cause us to develop chronic infections—e.g., chronic sinus infections, frequent cold sores, Epstein Barr virus, etc.[380] Immune suppression may also make us more likely to develop chronic fatigue, as discussed in Chapters 19-20. What makes matters worse is that immune suppression also makes it harder for doctors to diagnose infections because our immune systems are no longer behaving normally. For example, fever is an indicator of illness but we may not develop a fever during an infection because our immune system is suppressed.

It is also widely known that toxic chemicals affect our central nervous system.[381] As a result, the can disrupt both our mood and our mental health. Problems such as depression, anxiety, aggression, memory impairment, and even schizophrenia are fair game when it comes to toxic chemical exposure.[382] In children, problematic chemicals can set off a chain of reactions

---

[377] See e.g., Chang Gue Son, et al., "Complementary and Alternative Medicine for Diseases and Disorders in Digestive Tract: Basic to Clinics," *Evidence Based Complimentary Alternative Medicine*, 2013, 565279 (Oct 2013).

[378] Monica Feldman, et al., "Cultivate Your Probiotic Performance: Market Trends and Innovative Solutions," Lonza Whitepaper (2019), available at https://www.probiotaevent.com/wp-content/uploads/2019/01/Probiotics_Whitepaper_A4_10_2018_showpad. pdf; Aslam Shaikh, "Demand for Probiotics is Increasing Significantly," *Dairy Reporter* (April 6, 2018); Harvard Health Letter, "Gut reaction: A limited role for digestive enzyme supplements," Harvard Health Publishing at Harvard Medical School (March 2018), available at https://www.health.harvard.edu/staying-healthy/gut-reaction-a-limited-role-for-digestive-enzyme-supplements.

[379] Jamie C. DeWitt, Sarah J. Blossom & Laurel A. Schaider , "Exposure to per-fluoroalkyl and polyfluoroalkyl substances leads to immunotoxicity: epidemiological and toxicological evidence," *Journal of Exposure Science & Environmental Epidemiology* 29, 148–156 (March 2019); M. Chalubinski & M.L. Kowalski, "Endocrine disrupters – potential modulators of the immune system and allergic response," *European Journal of Allergy and Clinical Immunology*, 61(11), 1326-1335 (Nov 2006).

[380] Jaakkola MS, Yang L, Ieromnimon A, et al., "Office work exposures and respiratory and sick building syndrome symptoms," *Occupational and Environmental Medicine*, 64:178-184 (May 2007); Melinda J. Tarr, "Chemical Alteration of Host Susceptibility to Viral Infection," in Richard G. Olsen, et al. (eds), *Comparative Pathobiology of Viral Diseases, Vol. 1*, (CRC Press 2019); cf. Robert Luebke, "Immune Function, Immunotoxicity, and Resistance to Infection and Neoplasia," a case study abstract presented at Moving Upstream: A Workshop on Evaluating Adverse Upstream Endpoints for Improved Decision Making and Risk Assessment in Berkeley, California (May 16-17, 2007), and published in *Environ Health Perspectives* 116(11), 1568–1575 (Nov 2008).

[381] Nicholas Ashford & Claudia Miller, *Chemical Exposures: Low Levels and High Stakes (2nd ed)*, 27, 49, 208 (Wiley & Sons 1998).

[382] Rachel Carson, *Silent Spring*, 198 (The Riverside Press 1962)and Nicholas Ashford & Claudia Miller, *Chemical Exposures: Low Levels and High Stakes*, 140-142 (2nd ed) (Wiley & Sons 1998).

that alter neurodevelopment, causing chronic behavior issues and diseases such as autism.[383] Indeed, autism rates in our youngest generations have climbed at alarming rates.[384]

Toxic chemicals in a fetus and in young children are most problematic.[385] Researchers recognize that adults "can tolerate levels of pollution that devastate their offspring."[386] The fetus is still in development in the womb and exposure can lead to potentially permanent changes in the metabolism, organs, nervous system, immune system hormone balance, and other aspects of their bodies. These changes may frequently not arise to the level of obviously visible birth defects, but may cause life-long challenges.[387] Even exposure in young children can lead to relatively permanent changes, as the metabolism and neural pathways of a child are still in development. For example, asthma rates have jumped steadily.[388] And childhood chronic fatigue—formerly unheard of—has become a public policy issue.[389] Indeed, 30% of American children are living with some form of chronic illness today; an astounding figure.[390]

---

[383] Robert K. Naviaux, "Metabolic Features of Cell Danger Response," *Mitochondrion* 16, 7-17 (May 2014).

[384] "Autism Spectrum Disorder: Prevalence," US Centers for Disease Control and Prevention, available at https://www.cdc.gov/ncbddd/autism/data.html (last visited Oct. 22, 2019).

[385] "America's Children and the Environment (Third Edition)," EPA 240-R-13-001, US EPA (Jan. 2013), available at https://www.epa.gov/americaschildrenenvironment.

[386] Theo Colborn, et al., *Our Stolen Future*, 155 (Penguin Group 1997).

[387] See e.g., Philip J. Landrigan, et al., "Children's Health and the Environment: Public Health Issues and Challenges for Risk Assessment," *Environmental Health Perspectives* 112(2) 257-265 (Feb 2004) and Gayle C. Windham, et al., "Autism Spectrum Disorders in Relation to Distribution of Hazardous Air Pollutants in the San Francisco Bay Area," *Environmental Health Perspectives* 114(9), 1438-1444 (Sept. 2006).

[388] US Centers for Disease Control and Prevention, Asthma Prevalence Tables, available at https://www.cdc.gov/asthma/asthmadata.htm (last visited May 17, 2019).

[389] US Centers for Disease Control and Prevention, "ME/CFS in Children," available at https://www.cdc.gov/me-cfs/me-cfs-children/index.html (last visited Oct. 22, 2019), stating that up to 1 in 50 children suffer from chronic fatigue.

[390] "Metabolic Features of Myalgic Encephalomyelitis/Chronic Fatigue Syndrome," by Dr. Robert Naviaux, found in Transcript for the 9th CDC Myalgic Encephalomyelitis/Chronic Fatigue Syndrome Stakeholder Engagement and Communication (ME/CFS-SEC), May 25, 2017, available at https://www.cdc.gov/me-cfs/pdfs/transcript-naviaux-05272017.pdf

Chapter 16: What Our Bodies Do with Chemicals

As adults, we tend to accumulate toxic chemicals over our lifetime. Some are very hard to get rid of and can stay in our bodies for decades. As we grow older, we become increasingly prone to chronic illness as a result of their growing presence. Harmful chemicals enter our bodies primarily through our lungs, skin, or digestive system. We may breathe in fine particles in the air. Or absorb something we touched or bathed in. Or we may ingest something by directly putting it in our mouth. Or by putting our hands to our mouth or eyes after touching a contaminated surface.

Once synthetic chemicals are in our system, they generally will be transported into our blood stream.[391] Once these foreign substances are in our blood, our bodies seek to filter them out and eliminate them. We get into trouble when our bodies cannot eliminate the amount of problematic chemicals thrown at us quickly enough. We all eliminate foreign chemicals through our urine, our feces, sweat and our lungs. Unfortunately, it can take years and even decades to eliminate some of these chemicals.[392]

Women of childbearing age eliminate a significant amount of toxic substances through menstrual blood, breast milk, and the placenta.[393] This

---

[391] The exception are substances too big in their most essential form to be transported through the small capillaries leading into our blood vessels. One such example is asbestos, which are fine particles that will not break down once they enter our lungs. As a result, they get stuck in our lungs with potentially disastrous consequences.

[392] Jianjie Fu, et al., "Occurrence, temporal trends, and half-lives of perfluoroalkyl acids (PFAAs) in occupational workers in China," *Scientific Reports* 6 (Dec. 1, 2016); M. Stefanidou, et al., "Human Exposure to Endocrine Disruptors and Breast Milk," *Endocrine, Metabolic & Immune Disorders-Drug Targets*, 9(3), 269-276 (2009).

[393] Jianjie Fu, et al., "Occurrence, temporal trends, and half-lives of perfluoroalkyl acids (PFAAs) in occupational workers in China," *Scientific Reports* 6 (Dec. 1, 2016); M.

also means, however, that whatever synthetic substances a woman has in her body will be shared with her child--both during pregnancy and while breast feeding. As discussed in Chapter 15, this has the unfortunate effect of altering the physiology of even the unborn fetus.

Our liver is the primary organ tasked with detoxification. The liver initiates chemical processes that work to eliminate foreign substances from our bodies.[394] This can take days, weeks, months or years, depending on the chemical.[395] A number of problematic chemicals are known to damage our liver's ability to eliminate foreign substances—including certain pesticides.[396] A damaged liver will not be able to detoxify properly. Liver damage will thus lead to more accumulation and more chemical damage. For this reason, those of us who already have had a significant amount of chemical exposures and overloaded livers will be less tolerant to future chemical exposures and much more easily sickened by the problematic chemicals and substances thrown our way (see Chapters 19-21).

When our liver is not functioning properly many other problems can arise.

> [The liver] presides over so many vital activities that even the slightest damage to it is fraught with serious consequences. Not only does it provide bile for the digestion of fats, but...is deeply involved in the metabolism...It stores sugar in the form of glycogen and releases it as glucose in carefully measured quantities to keep the blood sugar at a normal level. It builds body proteins, including some essential elements of blood plasma concerned with blood-clotting. It maintains cholesterol at its proper level in the blood plasma, and inactivates the male and female hormones when they reach excessive levels.

> —Rachel Carson[397]

Stefanidou, et al., "Human Exposure to Endocrine Disruptors and Breast Milk," *Endocrine, Metabolic & Immune Disorders-Drug Targets*, 9(3), 269-276 (2009).

[394] DeAnn Liska, "The Detoxification Enzyme System," *Alternative Medicine Review*, 3(3), 187-198 (July 1998).

[395] Jianjie Fu, et al., "Occurrence, temporal trends, and half-lives of perfluoroalkyl acids (PFAAs) in occupational workers in China," *Scientific Reports* 6 (Dec. 1, 2016); M. Stefanidou, et al., "Human Exposure to Endocrine Disruptors and Breast Milk," *Endocrine, Metabolic & Immune Disorders-Drug Targets*, 9(3), 269-276 (2009)

[396] See e.g., Rachel Carson, *Silent Spring*, 191-192 (The Riverside Press 1962); Anthony Samsel and Stephanie Seneff, "Glyphosate's Suppression of Cytochrome P450 Enzymes and Amino Acid Biosynthesis by the Gut Microbiome: Pathways to Modern Diseases," *Entropy* 15(4), 1416-1463 (April 2013).

[397] Rachel Carson, *Silent Spring, 191* (The Riverside Press 1962).

Many of us have poorly functioning livers; indeed liver disease has climbed steadily.[398]

Our liver detoxification process also relies heavily on us having plenty of vitamins, minerals and nutrients at our disposal.[399] If we run short on any nutrients, detoxification of any foreign substances will be delayed. Unfortunately toxic chemicals can effectively deprive our bodies of these essential nutrients by interfering with our normal body chemistry.[400] Robert K. Naviaux, a researcher who studies chronic illness, has observed that human bodies respond to toxic chemical overload by shutting down, making it more difficult for us to absorb nutrients. Indeed, exposure to toxic chemicals can lead to a hibernation-like response in the body, meant to protect us from further exposure. This response has the unfortunate adverse effect of slowing down our ability to assimilate, utilize and/or produce any number of essential proteins, amino acids, and vitamins (including Vitamin D and certain B Vitamins).[401] Indeed, Vitamin D, Vitamin B-12, and other nutrient deficiencies are becoming an American epidemic.[402]

The more toxic our bodies become, the more nutrients we will need to keep up with the elimination of foreign substances. At this point in history, I would venture that our bodies need significantly more nutrients than we would ever be able to get from our diet in order to keep up with our toxic chemical world. Many of us are already buying vast quantities of supplements to help support our health. The market for supplements has grown to a $117 billion dollar industry in the last decades, with no signs of slowing down.[403] We are feeling the body burden of having to detox and looking for

---

[398] William N Hannah Jr. & Stephen A Harrison, "Noninvasive imaging methods to determine severity of nonalcoholic fatty liver disease and nonalcoholic steatohepatitis," *Hepatology* 64(6), 2234-2243 (June 24, 2016).

[399] Mark Percival, 'Phytonutrients & Detoxification," *Clinical Nutrition Insights*, 5(2) (Jan. 1997); CS Yang, et al., "Dietary effects on cytochromes P450, xenobiotic metabolism, and toxicity," *The FASEB Journal* 6(2), 737-744 (January 1992).

[400] See e.g., Anthony Samsel and Stephanie Seneff, "Glyphosate, pathways to modern diseases II: Celiac sprue and gluten intolerance," *Interdisciplinary Toxicology* 6(4), 159-184 (March 2014). For a technical explanation of how the P-450 liver enzymes are critical for the body's use of essential nutrients, see James P. Hardwick (ed.) *Cytochrome P450 Function and Pharmacological Roles in Inflammation and Cancer* (Elsevier 2015).

[401] Robert K. Naviaux, "Metabolic Features of Cell Danger Response," *Mitochondrion* 16, 7-17 (May 2014).

[402] Jordan Lite, "Vitamin D deficiency soars in the U.S., study says," *Scientific American* (March 23, 2009); Judy McBride, "B12 Deficiency May Be More Widespread Than Thought," *USDA Agricultural Research Service* (August 2, 2000); Victoria J. Drake, "Micronutrient Inadequacies in the US Population: an Overview," Oregon State University Linus Pauling Institute (November 2017).

[403] Tomislaw Mestrovic, "Nutraceutical Industry," *Medical & Life Sciences News* (Aug. 23, 2018).

solutions, even if we do not consciously understand what is going on inside of us.

When our bodies are overwhelmed with more chemicals than we can timely handle, a few things are likely to happen. First, the chemicals in our system start to adversely affect us in countless different ways. They may inhibit the normal functioning of our metabolism, organs, hormones, nervous system, and many other essential processes.[404] Our digestion and immune system are also likely to suffer because they are closely intertwined with our metabolism; indeed, metabolism is "the language the brain, gut and immune system use to communicate."[405]

Our bodies will then attempt to quarantine some amount of toxic chemicals in our fat cells.[406] The more toxic chemicals we put in, the more fat our bodies will try to generate in order to absorb toxic chemicals out of our blood stream to protect us. This is why some people may feel like despite their efforts to eat well and exercise regularly, they find themselves building an extra reserve of fat around their waist line.

If we are not successful at either eliminating chemicals from our bodies or tucking them away in body fat, then our immune system may try harder to get rid of them from our cells, tissues, and other places. The battle may feel to us like inflammation, fatigue, joint and muscle pain, irritability, or other symptoms."[407] If this lasts long enough, we may develop an autoimmune disease.[408] Indeed, autoimmune conditions have been directly associ-

---

[404] A.C. Gore, et al., "EDC-2: The Endocrine Society's Second Scientific Statement on Endocrine-Disrupting Chemicals," *Endocrine Review* 36(6): E1–E150 (Dec 2015).

[405] Robert K. Naviaux, "Metabolic Features of Cell Danger Response," *Mitochondrion* 16, 7-17 (May 2014).

[406] Michele La Merrill, et al., "Toxicological Function of Adipose Tissue: Focus on Persistent Organic Pollutants," *Environmental Health Perspectives* 121(2), 162-169 (Feb 2013).

[407] As described in *Silent Spring*, scientists who intentionally exposed themselves to notable amounts of DDT (a highly toxic pesticide banned years ago) described sensations of "tiredness," "heaviness," "muscle weakness," "joint pain" and "aching of the limbs." There was also "extreme irritability," "insomnia," "nervous tension" "feelings of acute anxiety" and "great distaste for work of any sort" and in more acute cases, physical tremors). Rachel Carson, *Silent Spring,* 193 (The Riverside Press 1962). By way of another example, acute exposure to mercury, will cause fever, fatigue, and other symptoms. Chronic long-term exposure to mercury will ultimately lead to neurological problems such as memory loss, irritability or depression. Martin G. Belson, et al., "Case Definitions for Chemical Poisoning," Division of Environmental Hazards and Health Effects, National Center for Environmental Health, 54(RR01);1-24 (January 14, 2005), available at https://www.cdc.gov/mmwr/preview/mmwrhtml/rr5401a1.htm.

[408] See e.g., M. Firoze Khan & Gagduo Wang, "Environmental agents, oxidative stress and autoimmunity," *Current Opinion in Toxicology"* 7, 22-27 (Feb. 2018) and Emenuela Corsini et al., "Steroid hormones, endocrine disrupting compounds and immunotoxicology," *Current Opinion in Toxicology* 10, 69-73 (Aug. 2018).

ated with toxic chemical exposure.[409] Allergies, asthma, psoriasis, rheuma-toid arthritis, fibromyalgia, lupus, inflammatory bowel disease, multiple sclerosis, type 1 diabetes, Hashimoto's thyroiditis, endometriosis, and many other conditions are a form of autoimmune response within the body.[410]

With each year and decade we better understand the impacts of toxic chemicals on our bodies. And yet we rarely hear about this growing health epidemic. Instead, we are told that our health issues are being caused by a sedentary lifestyle, stress, our diet, genetics or similar factors. Our self-imposed plague of chronic illness is now swallowing up three quarters of healthcare spending in the US.[411] And our life expectancy—for the first time in history—is now lower than in previous generations.[412]

---

[409] Annarosa Floreani,"Environmental Basis of Autoimmunity," *Clinical Reviews in Allergy & Immunology* 50:3, 287-300 (June 2016); Kenneth Michael Pollard, "Environment, autoanti-bodies, and autoimmunity," Frontiers in Immunology 6:60 (Feb 2015).

[410] "NIH scientists find link between allergic and autoimmune diseases in mouse study," National Institutes of Health News Release (June 2, 2013), available at https://www.nih.gov/news-events/news-releases/nih-scientists-find-link-between-allergic-autoimmune-diseases-mouse-study; "Autoimmune Disease List," American Autoimmune Related Diseases Association, available at https://www.aarda.org/diseaselist/ (last visited October 15, 2019)

[411] Wullianallur Raghupathi and Viju Raghupathi, "An Empirical Study of Chronic Diseases in the United States: A Visual Analytics Approach to Public Health" *International Journal of Environmental Research & Public Health* 15(3), 431 (March 2018).

[412] Swanson, N.L., Leu, A., Abrahamson, J. & Wallet, B. "Genetically engineered crops, glyphosate and the deterioration of health in the United States of America," *Journal of Organic Systems* 9(2), 6-37 (Jan 2014).

Chapter 17: Hormone Disruptors

Many synthetic chemicals are known to interfere with essential communications in our bodies. They are known as hormone or endocrine disruptors. They are similar to the chemicals that nature has created, but slightly different. Which is a problem: Hormone disruptors are chemicals that fool our bodies into believing that they belong there and send mixed messages in our hormone system, disrupting what happens inside of us.[413]

Such chemicals do not have to be in our bodies in significant quantities—rather, they can be problematic at very trace amounts. A few parts per billion are enough to change the course of history.[414] Timing of exposure matters the most. Timing is most critical for a human being during growth and development. This means that an unborn fetus, infant, and child will be much more susceptible to hormone disrupting chemicals than the adult.

---

[413] Hormone disruptors can mimic hormones and bind to hormone receptors, hijacking them. Thaddeus T. Schug, et al., "Minireview: Endocrine Disruptors: Past Lessons and Future Directions," *Molecular Endocrinology* 30(8), 833-847 (Aug 2016). They can also interfere what our genes are doing, activating and deactivating genes that communicate with our hormones, dictating our growth, development, health and wellbeing. "Endocrine Disruptors: from Scientific Evidence to Human Health Protection," 18 a study commissioned by the PETI Committee of the European Parliament (2019). Lister Hill National Center for Biomedical Communications, U.S. National Library of Medicine National Institutes of Health, Department of Health & Human Services, "Genetics Home Reference: Help Me Understand Genetics - How Genes Work (July 16, 2019), *available from* https://ghr.nlm.nih.gov/; National Institutes of Health US National Library of Medicine, "Genetics Home Reference: What are proteins and what do they do?," available at https://ghr.nlm.nih.gov/primer/howgeneswork/protein (last visited July 17, 2019).

[414] Theo Colborn, et al., *Our Stolen Future* (Penguin Group 1996) and European Union, "Endocrine Disruptors: from Scientific Evidence to Human Health Protection" 11, a study commissioned by the PETI Committee of the European Parliament (2019).

Indeed, hormone disruptors are most potent when we are in the womb, where the blueprint for who we will grow into by the time we are born has yet to be determined. For this reason, Theo Colborn, the grandmother of the science behind hormone disruptors, called such chemicals "hand me down poisons" because they are passed down from generation to generation and can harm future generations the most. Researchers have found that the children and even grandchildren of people exposed to hormone disruptors will be most affected.[415] This is because hormone disruptors interfere with the development of the unborn child; and some of the cells found in the unborn child will be used for the development of that child's own unborn children later in life.

Hormone disruptors can significantly undermine the health of the adult body as well. Unfortunately medical doctors generally do not have the capacity to tell whether hormone disruptors are affecting us. They can tell whether our hormones appear low or high in blood tests, but not whether their function has been disrupted within the greater system. Nevertheless, scientists are able to measure that the widespread use of hormone disruptors has led to an increase in reproductive health problems, cancer of the reproductive organs, thyroid disorders, mood disorders, obesity, diabetes, and problems such as ADHD, IQ loss and autism.[416]

We are inundated with hormone disruptors in our modern everyday lives. There are approximately 1000 synthetic hormone disruptors used in our economy that are recognized by the international community, and likely many more.[417] Even chemicals once thought to be weak hormone disruptors have proven to be highly potent.[418] Scientists began understanding the impacts of these chemicals on our health approximately 30 years ago and

---

[415] Theo Colborn, et al., *Our Stolen Future*, 26 (Penguin Group 1996); Jinjun Chen, "The Mechanism of Environmental Endocrine Disruptors (DEHP) Induces Epigenetic Transgenerational Inheritance of Cryptorchidism" *PLOS ONE* 10(6), e0126403 (June 2, 2015).

[416] Bergman, A, et al. (ed), "State of the Science of Endocrine Disrupting Chemicals," ISBN: 978-92-807-3274-0 (UNEP), World Health Organization, Inter-Organization Programme for the Sound Management of Chemicals (2012) and "Endocrine Disruptors: from Scientific Evidence to Human Health Protection" 20, PE 608.866, European Union Policy Department for Citizens' Rights and Constitutional Affairs (May 2019).

[417] Thaddeus T. Schug, et al., "Minireview: Endocrine Disruptors: Past Lessons and Future Directions," *Molecular Endocrinology*, 30(8), 833-847 (August 2016). In 2012, there were nearly 800 known hormone disruptors. Bergman, A, et al. (ed), "State of the Science of Endocrine Disrupting Chemicals" viii, ISBN: 978-92-807-3274-0 (UNEP), World Health Organization, Inter-Organization Programme for the Sound Management of Chemicals (2012).

[418] Theo Colborn, et al., *Our Stolen Future*, 74 (Penguin Group 1996).

this was after already nearly 40 years of scientific observation.[419] And yet we have done very little to regulate them today, for reasons discussed in Part 1 of this book.

This is a tremendous problem because hormones dictate so much of our body functions—much more than our gender and reproduction, which is what most people associate with the term "hormones." Scientists describe hormones as the chemical messengers in our bodies, signaling what to do and when to do it. They are, however, more than messengers—they are gate-keepers of what happens inside of us and can turn on/off biological activity.[420] At the direction of our genes, hormones are distributed in our bodies to hormone receptors with the aid of various proteins.[421] Once a hormone binds with a hormone receptor, a series of events can unfold in our body—ranging from temperature regulation to reproduction.

The following physiological functions are governed by our hormones:[422]

- Cellular growth & energy production
- Insulin & glucose regulation
- Muscle & organ function
- Fat metabolism & appetite
- Stomach acid production & digestion
- Fluid balance & blood pressure
- Blood cell production & oxygen delivery
- Development in the womb

Our hormones are produced by our thyroid, thymus, pineal, pituitary, hypothalamus, stomach, pancreas, liver, kidney, adrenals, parathyroid, and adipose tissue, as well as our reproductive organs.[423] These organs work together, coordinating various hormones. For example, the hypothalamus and pituitary glands in our brains work closely with our thyroid, adrenal

---

[419] See Bern, H et al., "Statement from the work session on chemically-induced alterations in sexual development: the wildlife/human connection" in Racine, Wisconsin (July 1991), republished in Theo Colborn, et al., *Our Stolen Future* (Penguin Group 1996).

[420] Theo Colborn, et al., *Our Stolen Future*, 71 (Penguin Group 1996).

[421] "Endocrine Disruptors: from Scientific Evidence to Human Health Protection" 23, PE 608.866, European Union Policy Department for Citizens' Rights and Constitutional Affairs (May 2019).

[422] Bergman, A, et al. (ed), "State of the Science of Endocrine Disrupting Chemicals" 5, ISBN: 978-92-807-3274-0 (UNEP), World Health Organization, Inter-Organization Programme for the Sound Management of Chemicals (2012).

[423] Bergman, A, et al. (ed), "State of the Science of Endocrine Disrupting Chemicals" 4, ISBN: 978-92-807-3274-0 (UNEP), World Health Organization, Inter-Organization Programme for the Sound Management of Chemicals (2012); Thaddeus T. Schug, et al., "Minireview: Endocrine Disruptors: Past Lessons and Future Directions," *Molecular Endocrinology*, 30(8), Pages 833–847 (August 2016).

glands, and reproductive organs to keep many of our body's physiological processes working properly.[424]

Just a single dose of a hormone disrupting chemical at the wrong time can have long-lasting life changes on these systems.[425] For example, estrogen at the right time can cause a fetus to grow properly. At a different time, it can cause a fetus to abort or can inhibit conception. Estrogen is an essential hormone in both men and women, one that can help both genders maintain a healthy body chemistry. But it can also unleash unbridled cell growth leading to cancer when hormone levels or ratios are not what they should be.[426] A number of man-made chemicals mimic or interfere with natural estrogen, causing problems throughout our life stages.

There is some manipulation of hormones in the natural world by life itself. For example, some hormone disrupting chemicals have evolved in plants. These substances are known as phytoestrogens and mimic reproductive hormones to protect the plant communities. When animals graze too much on plants containing phytoestrogens, they are ingesting a natural form of birth control.[427] Having overgrazed on the plants, these animals will not reproduce and there will be less predators in the future to graze on the plant. In response to such hormone disruptors, we have developed chemicals that circulate in our blood stream and prevent us from soaking up too many phytoestrogens.[428] For this reason, most of us will not become infertile from eating too much edamame, for example.

This is not the case with man-made hormone disruptors. We have introduced so many of these substances into our world in such a short period of time that life has not been able to adjust on an evolutionary level. Our bodies are not prepared to deal with, soak up, or defend themselves against synthetic hormones, as we have not had enough time on an evolutionary scale to do so.[429]

---

[424] "Endocrine Disruptors: from Scientific Evidence to Human Health Protection" 18, PE 608.866, European Union Policy Department for Citizens' Rights and Constitutional Affairs (May 2019).

[425] Theo Colborn, et al., *Our Stolen Future*, 112, 118-119 (Penguin Group 1996). For example, rats fed a trace dose of dioxin during the 15th day of pregnancy bore male offspring with a 50% reduction in sperm count. These same males also appeared to be feminized in their display of mating behavior.

[426] "Endocrine Disruptors: from Scientific Evidence to Human Health Protection" 18, PE 608.866, European Union Policy Department for Citizens' Rights and Constitutional Affairs (May 2019). "Help Me Understand Genetics - How Genes Work," in *Genetics Home Reference*, US National Library of Medicine (July 2019); "What are proteins and what do they do?" in *Genetics Home Reference*, US National Library of Medicine (July 2019).

[427] Theo Colborn, et al., *Our Stolen Future*, 76-78 (Penguin Group 1996).

[428] Theo Colborn, et al., *Our Stolen Future*, 73 (Penguin Group 1996).

[429] Theo Colborn, et al., *"Our Stolen Future*, 73, 81-82, 140 (Penguin Group 1996).

The most studied hormone disruptors have been shown to cause not one but many health problems.[430] And hormone disruptors also tend to be persistent, bio-accumulative, and toxic. As a result, they can build up in our bodies over time and across generations. We started exposing ourselves to hormone disruptors such as DDT, PCBs, dioxins as early as the 1940s. Although these particular chemicals have been restricted in commerce for some decades, they have left a legacy in our bodies and environment. They are passed down from mother to child and through our food chain. Even if we are able to eliminate them from our system, they return to the environment to be picked up by some other life form. They may also come back to haunt us when we drink water or eat food contaminated with them.

Unfortunately there are additional known hormone disruptors that have entered the marketplace in the last decades and which are actively used today. Many of them are also persistent, bio-accumulative, and toxic. Known hormone disruptors are commonly found in plastics (e.g., packaging, electronics), fragranced products (e.g., cleaning products, personal care products), pesticides, products containing PFAS (e.g., stain-resistant, waterproof and non-stick coatings on apparel, furniture) and flame retardants (used in electronics, as well as furniture made between 1975 and 2015).[431]

---

[430] A.C. Gore, et al., "EDC-2: The Endocrine Society's Second Scientific Statement on Endocrine-Disrupting Chemicals," Endocrine Review 36(6), E1–E150 (Dec 2015); "Endocrine Disruptors: from Scientific Evidence to Human Health Protection" 23, PE 608.866, European Union Policy Department for Citizens' Rights and Constitutional Affairs (May 2019); Theo Colborn, et al., *Our Stolen Future* (Penguin Group 1996). See also "International Cooperation: Persistent Organic Pollutants: A Global Issue, A Global Response," US EPA, available at https://www.epa.gov/international-cooperation/persistent-organic-pollutants-global-issue-global-response (last visited July 17, 2019).

[431] "Flame Retardants in Furniture," Green Science Policy Institute, available at https://greensciencepolicy.org/topics/furniture/ (last visited Feb. 4, 2020); Nicole C. Deziel, et al., "Exposure to Polybrominated Diphenyl Ethers and a Polybrominated Biphenyl and Risk of Thyroid Cancer in Women: Single and Multi-Pollutant Approaches," *Cancer Epidemiology, Biomarkers & Prevention* 28(20) 1755-1764 (Oct 2019); Thaddeus T. Schug, et al., "Minireview: Endocrine Disruptors: Past Lessons and Future Directions," *Molecular Endocrinology*, 30(8), 833–847 (August 2016); "Endocrine Disruptors: from Scientific Evidence to Human Health Protection" 11, PE 608.866, European Union Policy Department for Citizens' Rights and Constitutional Affairs (May 2019). See also Chapter 14 for plastics, Chapter 13for fragranced products, Chapters 8-10 for pesticides, and Chapters 11-12 for PFAS.

CHAPTER 18: DELAYED EFFECTS

What is most troubling about hormone disruptors is that in many cases their impact may not make themselves known until years later, and their effects are not very obvious to us. We do not drop dead immediately, but we may develop chronic health issues over our lifetime. Children may not be born with an extra arm, but their bodies may not develop to be as healthy as they were meant to be. And there are a number of ways that hormone disrupting chemicals are known to cause us problems:

Thyroid dysfunction.[432] In adults, the thyroid is responsible for regulating the metabolism, for bowel function, heart rate, body temperature, menstrual regularity, etc. It is a very important gland. Disruption of the thyroid hormones can lead to problems such as fatigue, weight gain, cold intolerance, hair loss, or chronic muscle pain, among other issues.[433] It can also lead to thyroid cancer.[434] Thyroid hormones will not necessarily appear low on blood tests if we are affected by hormone disruptors; it is enough that these chemicals are interfering with the proper functioning of the thyroid gland to make them problematic. They may have displaced natural thyroid hormones in the body or otherwise prevented naturally produced hormones from doing

---

[432] A.C. Gore, et al., "EDC-2: The Endocrine Society's Second Scientific Statement on Endocrine-Disrupting Chemicals," *Endocrine Review* 36(6): E1–E150 (Dec 2015).

[433] "For Patients - Thyroid FAQs" UCLA Endocrine Center UCLA Health, available at https://www.uclahealth.org/Endocrine-Center/thyroid-faqs (last visited August 27, 2019).

[434] "Endocrine Disruptors: from Scientific Evidence to Human Health Protection" 25, PE 608.866, European Union Policy Department for Citizens' Rights and Constitutional Affairs (May 2019).

their job.[435] Hormone disruptors do appear to be affecting our thyroids. We have a significant thyroid disease epidemic in modern society, with problems ranging from hypothyroidism to thyroid cancer.[436] Even house pets are now affected by these health problems.[437]

Brain Development.[438] In the fetus, thyroid hormones are responsible for development of the brain and nervous system. Thyroid hormones play an important role in getting nerve cells to develop in a healthy way and take their rightful place in the make-up of the human brain.[439] The fetus is entirely reliant on the mother's thyroid hormone for its own brain development. Thyroid hormone disruption in the mother will disturb proper fetal development of the brain of her unborn child. This can lead to a variety of neurodevelopment disorders, such as lower IQ, behavioral issues, anxiety, and ADHD, for example.[440] All of these disorders are on the rise.[441]

Type 2 Diabetes.[442] Insulin is a hormone that is needed to feed the cells with glucose from the blood stream for purposes of energy production. Many hormone disruptors can cause insulin resistance or type 2 diabetes;

[435] Theo Colborn, et al., *Our Stolen Future*, 187 (Penguin Group 1996).

[436] Lindsey Enewold, et al., "Rising Thyroid Cancer Incidence in the United States by Demographic and Tumor Characteristics, 1980-2005," *Cancer Epidemiology, Biomarkers & Prevention*, 18(3), 784-791 (March 2009); Marta Benedetti et al, "Incidence of Breast, Prostate, Testicular, and Thyroid Cancer in Italian Contaminated Sites with Presence of Substances with Endocrine Disrupting Properties," *International Journal of Environmental Resources and Public Health* 14(4), 355-366 (March 2017); Miranda M. Fidler et al, "A global view on cancer incidence and national levels of the human development index," *International Journal of Cancer*, 139 2436-2446 (Aug. 2016).

[437] Mark Peterson, "Hyperthyroidism in Cats: What's causing this epidemic of thyroid disease and can we prevent it?," *Journal of Feline Medicine and Surgery* 14(11), 804-818 (October 2012). See also W Jean Dodds & Diana R Laverdure, *The Canine Thyroid Epidemic*, (Dogwise Publishing 2011).

[438] A.C. Gore, et al., "EDC-2: The Endocrine Society's Second Scientific Statement on Endocrine-Disrupting Chemicals," *Endocrine Review*, 36(6): E1–E150 (Dec 2015).

[439] Theo Colborn, et al., *Our Stolen Future*, 187 (Penguin Group 1996).

[440] "Endocrine Disruptors: from Scientific Evidence to Human Health Protection" 20-21, 32, PE 608.866, European Union Policy Department for Citizens' Rights and Constitutional Affairs (May 2019).

[441] See Teresa M. Attina, et al., "Exposure to endocrine-disrupting chemicals in the USA: a population-based disease burden and cost analysis," *The Lancet Diabetes & Endocrinology* 4(12), 993-1003 (December 2016) (quantifying loss of IQ); "ADHD diagnosis throughout the years: Estimates from published nationally representative survey data," Centers for Disease Control (last visited February 5, 2020); Jenny Marie, "Millennials and Mental Health," National Alliance on Mental Illness (Feb. 27, 2019) (anxiety); Mark Townsend, "Massive rise in disruptive behaviour, warn teachers," *The Guardian* (March 23, 2013); "Rise in Aggressive Student Behavior in the Classroom Impacting Students, Teachers," Connecticut Education Association Statement (March 14, 2018).

[442] A.C. Gore, et al., "EDC-2: The Endocrine Society's Second Scientific Statement on Endocrine-Disrupting Chemicals," *Endocrine Review* 36(6): E1–E150 (Dec 2015); see also Wang, et al., "Treating Type 2 Diabetes Mellitus with Traditional Chinese and Indian Medicinal Herbs," *Evidence Based Complementary Alternative Medicine* (May 2013).

for this reason, they are called diabetogens. If the insulin receptors are inter-fered with, a person's body will be unable to bring glucose into the cells. Energy levels will suffer and rising glucose levels in the blood stream will cause damage throughout the body. If the problem becomes serious enough, the person will develop Type 2 diabetes. Diabetes has grown steadily in the developed world and is linked to hormone-disrupting chemical exposure.[443]

Weight gain.[444] Research has also shown that certain hormone disrup-tors predispose people to becoming fatter. Such chemicals are known as obesogens. If you have ever taken an estrogen-based birth control pill, you may have learned that it can make you put on weight fairly quickly. Hormone disruptors predispose people to obesity the most when the exposure takes place before birth or early on in life. They can still act upon people later in life, however, causing them to gain weight. More fat receptors may develop as a result of exposure, and these chemicals can actually signal the body to produce fat at unusually high rates. Obesity—including in children—has become an epidemic in modern society and scientists believe hormone disruptors are playing a contributory role.[445] My personal observation is that women who are regularly exposed to hormone disruptors on the job (those who work with pesticides, plastics and fragranced cleaning products in particular) tend to hold excessive fat around their stomachs.[446] Another personal observation from living and traveling abroad is that people are the fattest in countries where plastics (see Chapter 22) have been most perva-sively integrated into daily life.

Chemical Sensitization. Exposure to hormone disrupting chemicals in the womb, after birth, and even later in life can lead to sensitization to toxic chemicals through a process known as hormone imprinting. During hormone imprinting, newly divided cells in the embryo or even in the adult human being develop their receptivity to hormones based on what they find in their environment at the time.[447] For example, exposure to synthetic estrogens at the beginning stages of life can lead to an increased sensitivity to estrogen.

---

[443] Marcelo G. Bonini & Robert M. Sargis, "Environmental toxicant exposures and type 2 diabetes mellitus: Two interrelated public health problems on the rise," *Current Opinion in Toxicology* 7, 52-59 (Feb. 2018).

[444] A.C. Gore, et al., "EDC-2: The Endocrine Society's Second Scientific Statement on Endocrine-Disrupting Chemicals," *Endocrine Review* 36(6): E1–E150 (Dec 2015).

[445] Jin Taek Kim, et al., "Childhood obesity and endocrine disrupting chemicals," *Annals of Pediatric Endocrinology & Metabolism* 22(4), 219-225 (Dec 2017); and Jerrold J. Heindel, "Endocrine Disruptors and the Obesity Epidemic," *Toxicological Sciences* 76(2) 247–249 (Dec 2003).

[446] Pesticides, plastics, and fragranced products are discussed in various chapters of this book.

[447] G. Csaba, "The biological basis and clinical significance of hormonal imprinting, an epigenetic process," *Clinical Epigenetics* 2(2), 187–196 (Aug 2011).

This leads to an increased risk of cancer later in life in individuals who have become more sensitive to estrogen, because estrogen can cause cells to grow very quickly.[448] Various forms of chemical sensitization and even desensitization can occur during the hormone imprinting process.[449] For example, we may become more insulin resistant and intolerant of sugar as a result of hormone imprinting. Scientists also suspect that chemical sensitization is the culprit for the rise in hormone-responsive cancers, such as breast, prostrate, ovary, and uterus cancers.[450]

Immune System Problems.[451] Studies on hormone disrupting chemicals have shown that they adversely affect the body's immune system. These impacts are compounded as chemicals accumulate in our bodies over time. As a result, we are more likely to have a suppressed immune system, allergic reactions, chronic inflammation, and autoimmune problems as we age.[452] At the same time, the immune system is most vulnerable to long-term disruptions if toxic chemical exposure occurs in the womb.[453] Indeed, vaccination studies have shown that children exposed to certain hormone disruptors (likely even before birth) were unable to produce antibodies to a number of viruses.[454]

Reduced Ability to Deal with Stress.[455] Hormone disruptors target the ability to produce proper stress hormones in the adrenal glands. Babies exposed to hormone disruptors are known to be much more cranky and less resilient to future disturbances than those who have not suffered such exposure, for example. Even adults who have been exposed to stress-producing hormone disruptors may be fine, so long as life is good, but when stress

---

[448] Theo Colborn, et al., *Our Stolen Future*, 62-63, 122-131, 183 (Penguin Group 1996).

[449] G. Csaba, "The biological basis and clinical significance of hormonal imprinting, an epigenetic process," *Clinical Epigenetics* 2(2), 187–196 (Aug 2011); see also Theo Colborn, et al., *Our Stolen Future*, 79-80 (Penguin Group 1996).

[450] Theo Colborn, et al., *Our Stolen Future*, 172 (Penguin Group 1996); see also Marta Benedetti et al, "Incidence of Breast, Prostate, Testicular, and Thyroid Cancer in Italian Contaminated Sites with Presence of Substances with Endocrine Disrupting Properties," *International Journal of Environmental Resources and Public Health* 14(4), 355 (March 2017); Miranda M. Fidler et al, "A global view on cancer incidence and national levels of the human development index," *International Journal of Cancer*, 139, 2436-2446 (Aug. 2016).

[451] A.C. Gore, et al., "EDC-2: The Endocrine Society's Second Scientific Statement on Endocrine-Disrupting Chemicals," *Endocrine Review* 36(6): E1–E150 (Dec 2015).

[452] Theo Colborn, et al., *Our Stolen Future*, 63, 88-89, 144, 147, 161-163 (Penguin Group 1996); and M. Chalubinski & M.L. Kowalski, "Endocrine disrupters – potential modulators of the immune system and allergic response," *European Journal of Allergy and Clinical Immunology* 61(11), 1326-1335 (November 2006).

[453] Theo Colborn, et al., *Our Stolen Future*, 161-162 (Penguin Group 1996).

[454] Our Stolen Future at 107 and M. Chalubinski & M.L. Kowalski, "Endocrine disrupters – potential modulators of the immune system and allergic response," European Journal of Allergy and Clinical Immunology, Vol. 61, Issue 11 (November 2006) at 1326-1335.

[455] Theo Colborn, et al., *Our Stolen Future*, 85, 114, 137, 192-194 (Penguin Group 1996).

enters the picture, however—they can be very much affected. They may overreact to even small problems and potentially have extreme difficulties in more stressful situations. We already appear to be seeing the effects of stress-inducing hormone disruptors in the general population. Millennials (today's young adults) are much more anxious and stressed than those in generations before them.[456]

---

[456] Jenny Marie, "Millennials and Mental Health," National Alliance on Mental Illness (Feb. 27, 2019).

Chapter 19: Sex and Reproduction

Hormone disruptors are also causing changes in human fertility and reproductive organs. Many of these changes are being programmed into our bodies before we are even born.[457] In females, hormone disruptors are associated with the early onset of puberty; decreased fertility; irregular cycles; early menopause; fibroids; endometriosis; pelvic inflammation; abnormalities in the shape and structure of the vagina; uterus; and ovaries; challenges with breastfeeding, and complications during pregnancy and birth.[458]

In males, hormone disruptors are associated with reduced testosterone levels; reduced sperm count; malformed sperm; smaller penis and testes; malformations of the penis; reduced erections; cysts, lesions, chronic inflammation and/or hemorrhage of the reproductive organs; and reduced analgenital distance.

Many of these reproductive changes have not been formally tracked in the general population, but sperm counts have been studied extensively. Sperm counts have gone down in the Western world from 113 million per milliliter of semen in 1950 to approximately 47 million per milliliter of semen in 2011.[459]

---

[457] A.C. Gore, et al., "EDC-2: The Endocrine Society's Second Scientific Statement on Endocrine-Disrupting Chemicals," *Endocrine Review* 36(6): E1–E150 (Dec 2015).

[458] A.C. Gore, et al., "EDC-2: The Endocrine Society's Second Scientific Statement on Endocrine-Disrupting Chemicals," *Endocrine Review* 36(6): E1–E150 & Table 4 (Dec 2015).

[459] Compare. R. Sharpe & N. Skakkebaek, "Are Oestrogens Involved in Falling Sperm Counts and Disorders of the Male Reproductive Tract?" *The Lancet* 341, 1392-1395 (May 1993), cited in Theo Colborn, et al., *Our Stolen Future*, 173 (Penguin Group 1996) and H. Levine, et al., "Temporal trends in sperm count: a systematic review and meta-regression analysis," *Human Reproduction Update*, 23(6), 646–659 (Nov-Dec 2017).

Hormone disruptors can also alter the development of our gender and sexual identity if we are exposed to them at just the wrong time before birth.[460] While the X and Y chromosomes we all learned about in school are a starting point, our gender development cannot take place without hormones. By default, all mammals (including human beings) are programmed to develop as female in coordination with hormones that naturally affect our bodies in the womb. Without the unique intervention of additional hormones at critical stages of fetal development, we will not develop male organs or body chemistry, even if our genes dictate that we should be male.[461]

Transgender and hermaphrodite individuals (aka intersex) are those who likely started out on the road of becoming male, but did not fully complete the process because of some significant hormone disruption.[462] As a result, they may have sexual organs of both genders or the body chemistry of one gender but the sexual organs of another. Similarly, whether a person will be attracted to the opposite or the same sex is determined by their particular hormone chemistry before birth, which could possibly be altered by hormone disruptors.[463]

Before we ever understood the role of gender hormones in humans, scientists observed gender hormone disruption in wildlife living in polluted environments. Fishermen began observing intersex fish some decades ago. In response, scientists investigated and found that commercially synthesized hormone disruptors were the cause.[464] In North America, intersex whales and intersex turtles have also been documented as a result of exposure to hormone disruptors for some time.[465] We are now beginning to see these same types of hormonally induced changes in a critical mass of the human population.

Although the US government historically has not officially kept track of transgender, intersex, and gender neutral individuals, the evidence strongly suggests that these populations are notably on the rise. According to a recent Harris Poll survey, 20% of millennials identify as LGBTQ; this is a two-fold increase from prior generations recently surveyed.[466] The report continues on to say:

---

[460] A.C. Gore, et al., "EDC-2: The Endocrine Society's Second Scientific Statement on Endocrine-Disrupting Chemicals," *Endocrine Review* 36(6): E1–E150 (Dec 2015).
[461] Theo Colborn, et al., *Our Stolen Future*, 42-46 (Penguin Group 1996).
[462] Theo Colborn, et al., *Our Stolen Future*, 83 (Penguin Group 1996).
[463] Theo Colborn, et al., *Our Stolen Future*, 64-65 (Penguin Group 1996).
[464] Theo Colborn, et al., *Our Stolen Future*, 131-135 (Penguin Group 1996).
[465] Theo Colborn, et al., *Our Stolen Future*, 141-147, 152-153 (Penguin Group 1996).
[466] GLAAD, "Accelerating Acceptance," Harris Poll (2017).

Perhaps most striking, 12% of Millennials identify as transgender or gender nonconforming, meaning they do not identify with the sex they were assigned at birth or their gender expression is different from conventional expectations of masculinity and femininity—doubling the number of transgender and gender nonconforming people reported by Generation X.[467]

This has become a sufficiently common issue that many companies, universities, and other institutions have begun to incorporate a gender-neutral pronoun into their lexicon. The options or no longer he or she, but rather, he, she or ze. In some organizations, new employees are asked to specify for others whether they prefer to be called he, she or ze.[468]

Another interesting indicator that the rates of transgender, intersex, and gender neutral individuals is growing comes from Mattel, the maker of the improbably well-endowed Barbie dolls. In its most recent market strategy, Mattel has revealed a new line of gender-neutral Barbie dolls called Creatable World.

[These dolls] don't have broad shoulders like Ken or breasts like Barbie. Creatable World dolls can be played with as a boy, a girl, or neither. They're slim with androgynous faces and short hair and can be fitted with wigs and a wardrobe consisting of sneakers, graphic tees, hoodies, tutus, or camouflage pants.[469]

Androgynous is becoming the new normal.

A further alarming trend is that a significant number of millennials (young adults born between 1981 and 1996) who should be in their peak years of sexuality have a very limited sex drive or no sex drive at all.[470] These are sharp sexual hormone aberrations not present in recorded history. And this detour likely began in the womb before millennials were even born. One explanation provided by the media for millennials' lack of sex drive is lack of sex education. If you were born before the 1980s, you will be skeptical of this explanation.

---

[467] GLAAD, "Accelerating Acceptance," Harris Poll (2017).

[468] Eugene Volokh, "You can be fined for not calling people 'ze' or 'hir,' if that's the pronoun they demand that you use," *The Washington Post* (May 17, 2016); Kristin Tauras, "Which Pronoun Fits Your Employee?" McKenna Storer Law (June 18, 2019), available at https://www.mckenna-law.com/blog/which-pronoun-fits-your-employee/.

[469] Todd Perry, "Step Aside, Barbie. Mattel is launching its first line of gender-neutral dolls." *Upworthy* (Sept. 25, 2019), available at https://www.upworthy.com/barbies-creators-are-launching-their-first-line-of-gender-neutral-dolls?fbclid=IwAR23_Oyg2npRX0Hea4Xj XEO4iuqWFZe9RUKZtwDCA-tjTPtddUZPSISwhOI

[470] Kate Julian, "Why Are Young People Having So Little Sex?" *The Atlantic* (December 2018); Kelsey Borresen, "Here's What Millenials Say About Sex in Therapy," *HuffPost* (May 17, 2019).

CHAPTER 20: FUEL TO THE FIRE

Chemical mixtures are known to have "synergistic effects." This means that a given chemical can become even more toxic when another problematic chemical gets involved. For example, glyphosate becomes much more toxic when combined with the surfactant that is in the chemical formulation of RoundUp, as discussed in Chapter 10. Pyrethroids and organophosphates—two varieties of pesticides—also become even more toxic when combined together.[471]

Chemicals mix not only in product bottles, but also in the environment and even in our own bodies. They can bind together to form new molecules, or simply work in tandem to undermine normal physiological processes. One chemical can do preliminary damage to our system, inviting the next chemical to finish the job. There are also metabolites of chemicals—the breakdown products of these foreign substances. Break-down products can be even more toxic than the original substance and cause damage in the body before we are able to eliminate them.[472]

Because of these compounding effects, the sequence in which one is exposed to different chemicals can mean the difference between dodging a bullet and getting sick. Different chemicals have different life-spans within the human body. Some may take days to process and eliminate, others may

---

[471] E.G., Valliantanos with M. Jenkins, *Poison Spring*, 39 (Bloomsbury Press 2014).

[472] See e.g., "What You Know Can Help You - An Introduction to Toxic Substances," New York State Department of Health, available at https://www.health.ny.gov/environmental/chemicals/toxic_substances.htm (last visited February 1, 2020); and Anne-Christine Macherey & Patrick M. Dansette, "Chapter 25 - Biotransformations Leading to Toxic Metabolites: Chemical Aspects" in *The Practice of Medicinal Chemistry (Fourth Edition)*, Academic Press (2015).

take years. A long-lived chemical can lie relatively dormant inside of us, and get activated by the introduction of another problematic chemical weeks, months, or even years later.

> You can be exposed to chemical A and nothing happens. You can be exposed to chemical B and nothing happens. You can be exposed to chemical A and then B, nothing happens. But if you get exposed to chemical B and then A—boom—you are at risk for cancer.
>
> —Anonymous Industry Toxicologist

The trouble is that we do not really know what the chemical As and Bs are in this world. Nor do we fully know how much each of us is being exposed, and in what order. At the same time, it is very difficult to live in modern society and avoid exposure to toxic chemicals. There are people who spend a significant part of their energies trying to avoid additional chemical exposures. These are usually victims of environmental illnesses such as chronic fatigue, fibromyalgia and multiple chemical sensitivity (Chapters 19–21). And even they can tell you that it is nearly impossible to navigate the modern world without inadvertently being exposed to toxic chemicals.

Chemical compounding effects also occur when chemicals and germs get together.[473] Indeed, the viruses, bacteria and other microorganisms we are trying to avoid can become more dangerous to us under the influence of harmful chemicals. The chemistry that makes up a germ's environment can cause a relatively dormant and benign organism to go on the offensive and become a raging infection.[474] The mechanism by which this happens is known as "quorum sensing."[475]

Quorum sensing is a chemical process that causes microorganisms to swarm together[476] and actively start producing toxins, biofilms, spores, and generally attack their host in a really bad way.[477] Microorganisms decide to set up shop under certain chemical and environmental conditions; if those

---

[473] Nils-Gunnar Ilbäck & Göran Friman, "Interactions Among Infections, Nutrients and Xenobiotics," *Journal of Critical Reviews in Food Science & Nutrition* 47(5), 499-519 (June 2007); JP Desforges, et al., "Environmental contaminant mixtures modulate in vitro influenza infection," *Science of the Total Environment* 634, 20-28 (Sept 2018).

[474] John S. Parkinson, et al., "Signal Transduction Schemes of Bacteria," *Cell* 73, 857-871 (June 1993).

[475] Nandhitha Venkatesh and Nancy P. Keller, "Mycotoxins in Conversation with Bacteria and Fungi," *Frontiers in Microbiology* 10, 403 (March 2019).

[476] Donping Wang, et al., "Transcriptome profiling reveals links between ParS/ParR, MexEF-OprN, and quorum sensing in the regulation of adaptation and virulence in Pseudomonas aeruginosa," *BMC Genomics* 14, 618 (Sept 13, 2013).

[477] Kayeen Vadakkan, "Quorum sensing intervened bacterial signaling: Pursuit of its cognizance and repression," *Journal of Genetic Engineering and Biotechnology* 16:2, 239-252 (Dec 2018).

conditions are absent, they will lay relatively dormant in our ambient environment and even in our bodies. If these conditions are present, they will have a hay day.

Normally quorum sensing occurs when there is a critical mass of bacteria, viruses, fungi, etc., in one suitable place—when enough gather in a cozy environment, they start to emit chemical signals to each other, and that causes these microorganisms to attack us.[478] For example, cats have some dangerous bacteria in their mouths, but even if they lick our skin or we ingest their saliva accidentally, we do not get sick. The germs in cat saliva lie dormant in the ambient environment. It is only once we are bitten and enough bacteria begin to circulate in our system, and an attractive environment, that an infection can develop.[479]

Human bodies produce their own chemistry in order to resist viruses, bacteria, and other microorganisms.[480] Toxic chemicals can disrupt their ability to do this, however, leaving us vulnerable to illness.[481] COVID-19 has demonstrated the problematic relationship between chemicals and germs. For example, researchers have found that people with chronic health conditions caused by air pollution are at increased risk of dying from COVID-19.[482] Indeed, problematic chemicals can alter our internal or external environments in a way that encourage such infections. Reyes Syndrome, for example, occurs when the flu virus is made deadly to children because they have been exposed to certain problematic chemicals.[483] The pesticide residues found in honey can also potentially turn otherwise ambient levels of botulinum toxin (produced by bacteria) into deadly agents that can kill an infant.[484] Even

---

[478] Sajad Ahmad Padder et al, "Quorum sensing: A less known mode of communication among fungi," *Microbiological Research* 210, 51-58 (May 2018).

[479] Frederick A. Abrahamian and Ellie J. C. Goldstein, "Microbiology of Animal Bite Wound Infections," *Clinical Microbiology Review* 24(2), 231–246 (April 2011).

[480] Angelika Holm & Elena Vikstrom, "Quorum sensing communication between bacteria and human cells: signals, targets, and functions," *Frontiers in Plant Science* 5, 309 (June 2014).

[481] John S. Parkinson, et al., "Signal Transduction Schemes of Bacteria," *Cell* 73, 857-871 (June 1993).

[482] Xiao Wu, et al., "Exposure to air pollution and COVID-19 mortality in the United States: A nationwide cross-sectional study," *MedRxiv* (preprint rev. April 24, 2020). Individuals with diabetes—an endocrine system disorder linked to toxic chemicals (Chapter 18)—are at very high risk with COVID-19 as well. M. Puig-Domingo, et al., "*COVID-19 and endocrine diseases. A statement from the European Society of Endocrinology,*" *Endocrine* 68(1), 2-5 (April 2020).

[483] JF Crocker, et al.. "Biochemical and morphological characteristics of a mouse model of Reye's syndrome induced by the interaction of influenza B virus and a chemical emulsifier," *Laboratory Investigation* 54(1), 32-40 (Jan 1986).

[484] E.G., Valliantanos with M. Jenkins, *Poison Spring*, 92 (Bloomsbury Press 2014).

prescription drugs can activate germs. For example, certain diabetic drugs are known to increase the risk for urinary tract infections.[485]

Toxic chemicals can also cause us to develop adverse reactions to substances that are otherwise benign in our environment. The substance has to only be chemically similar enough to something toxic that our body begins to treat it as a threat. For example, chronic exposure to diesel exhaust has long been suspected of causing a sharp rise in cedar allergies in Japan.[486] Cedar, along with conifers and many other trees, are known to produce terpenes.[487] Terpenes are volatile organic compounds with similar chemical structures to toxic chemicals found in diesel exhaust.[488] The body likely begins to learn that diesel exhaust has a toxic effect on it, and activates an immune response to substances with similar chemical profiles.

Another common trigger for people with existing chemical exposures is mold. Mold itself can emit chemicals that are toxic to us, known as mycotoxins. However, the problem of mold toxicity is likely made worse by the toxic chemicals that have been introduced into the environment. Mold—just like any other living creature—has to metabolize all of the toxic chemicals around us. As a result, mold appears to be excreting our environmental pollution back at us in, including potentially toxic metabolites from the original problematic substances.[489]

A combination of mold, a viral infection and toxic chemicals is suspected to have led to a massive outbreak of chronic fatigue in the 1980s in the Lake Tahoe region.[490] According to Erik Johnson, a victim of the outbreak, there were a lot of adverse factors going for the local residents who fell ill with the disease. It started with a flu-like viral infection brought back from China by

---

[485] Michelle Lamb, "Urinary Tract Infections: Causes and Treatment Update," *US Pharmacist* 41(4), 18-21 (April 2016).
[486] Nicholas Ashford & Claudia Miller, *Chemical Exposures: Low Levels and High Stakes (2nd ed.)*, 63 (Wiley & Sons 1998).
[487] Mitsuyoshi Yatagai & Toshia Sato, "Terpenes of Leaf Oils from Conifers," *Biochemical Systematics & Ecology* 14(5), 469-478 (Oct. 1986); and Ben-Erik van Wyk & Michael Wink, *Phytomedicines, Herbal Drugs, and Poisons*, 50-63 Kew Publishing (2014).
[488] B.T. Jobson, et al., "On-line analysis of organic compounds in diesel exhaust using a proton transfer reaction mass spectrometer (PTR-MS)," *International Journal of Mass Spectrometry* 245(1-3), 78-79 (Aug 2005); "IARC: Diesel Engine Exhaust Carcinogenic," Press Release No. 213, International Agency for Research on Cancer (June 12, 2012).
[489] Ken Wilkins, et al., "Volatile metabolites from mold growth on building materials and synthetic media," *Chemosphere* 41(3), 437-446 (Aug 2000); B. Mousavi, et al., "Aspergillus species in indoor environments and their possible occupational and public health hazards," *Current Medical Mycology* 2(1), 36-42 (March 2016).
[490] Robert Steinbrook, "160 Victims at Lake Tahoe : Chronic Flu-Like Illness a Medical Mystery Story," *Los Angeles Times* (June 7, 1986); and Erik Johnson (personal communication, Feb. 3, 2020).

a marathon runner. The flu morphed from fleeting to persistent when other factors became involved.

Johnson observed that a notable subset of those who fell ill taught at a local high school that had sick building issues. The newly constructed facility was suspected of mold contamination and chemical off gassing. In addition, there was an unreported industrial solvent spill in town at the time from a small commercial tanker truck. Moreover, an unusual incidence of smog pollution plagued the area, potentially the cause of a massive die-off of algae in Lake Tahoe waters and visible damage to the surrounding forests. Each of these factors could have triggered a host of immune responses in those who were exposed to them. Combined, these events created an immune system overload in a portion of the local population, causing an outbreak of chronic illness.

CHAPTER 21: ENVIRONMENTAL ILLNESS

The term "environmental illness" was coined by a subset of medical doctors in the 1960s to describe illnesses known to be caused by environmental pollution.[491] A small band of doctors who called themselves clinical ecologists broke away from mainstream medicine and began treating patients for environmental illnesses using detoxification and desensitization techniques they developed out of allergy medicine and from experience in the field.[492]

Their findings regarding how the human body responds to environmental pollutants are a wealth of knowledge to us. It turns out that we respond to toxic chemicals in ways that are very similar to how we respond to smoking cigarettes, drinking alcohol, and enjoying caffeine. If we are not used to the amount we have taken in, we will have an adverse reaction. If we continue the habit at a manageable level, we will become used to the substance and our unpleasant symptoms will go away.[493]

Not only will the adverse effects go away, but we may start to feel peppy. Toxic chemicals we have adapted to—just like caffeine, nicotine, and

---

[491] David H. Nelson, "Clinical Ecology—Transforming 21st-Century Medicine with Planetary Health in Mind," *Challenges* 10(1), 15 (Feb 2019).

[492] David H. Nelson, "Clinical Ecology—Transforming 21st-Century Medicine with Planetary Health in Mind," *Challenges* 10(1), 15 (Feb 2019). A number of people I connected with in researching for this book have received treatment from clinical ecologists for environmental illness. The most famous of such clinics might be the Environmental Health Center in Dallas, Texas, which was founded by Dr. William Rea. Other individuals have sought chemical detoxification advice from environmental illness advocacy groups such as the Rocky Mountain Environmental Health Association.

[493] Nicholas Ashford & Claudia Miller, *Chemical Exposures: Low Levels and High Stakes (2nd ed.)*, ch 2 (Wiley & Sons 1998).

alcohol—can make us feel good. A whole host of chemical processes mobilize in the body as a result of low-level toxic chemical exposure that give us the perception that we are more alert, motivated, productive, witty or just positively stimulated.

If we amp up the dose too much we will go from feeling good to feeling not so good: we can become tense, jittery, easily angered, flush, nauseated, or get the chills, for example. If we keep amping up the dose to levels we cannot handle, we may become manic, catatonic, or even develop seizures.[494]

At low levels, we may become addicted to the toxic chemicals themselves—in the same way we can become addicted to nicotine, caffeine, or alcohol. We get addicted to the chemicals in paint fumes and fragrance and other nasty things if we expose ourselves regularly to them at levels our nervous system can handle.[495]

> I had heavy exposure to acetone at a job I had when I was 19 (in the 1970s). We used to wash our hands in it—we were working with resin and some kind of catalyst (I forget the name of it) but it produced a plastic and we made fake Tiffany lamps, and would wash the resin/plastic stuff off our hands with acetone. We also breathed in the fumes all day long - I was always a little high after work....

> —Anonymous Chronic Fatigue Sufferer

What happens when people become addicted? They experience withdrawal symptoms if they are away from their favorite drug. Mild withdrawal can feel like an allergic reaction—stuffy nose, rash, watery eyes, diarrhea, etc. More severe withdrawal can feel like fatigue, depression, muscle/joint pains, headache, and irregular heartbeat. Even more severe withdrawal can feel like confusion, apathy, inability to concentrate. At the most severe, a person may go into a complete stupor, become delusional, and even hallucinate.[496]

This addiction/withdrawal dynamic has long been recognized by occupational health specialists, who found that workers experience withdrawal symptoms over the weekend from being away from the chemicals they get exposed to at work. In certain industries—such as dynamite production— employees are actually advised to take grains of dynamite home and put them under their pillow so that they do not experience unpleasant with-

---

[494] Nicholas Ashford & Claudia Miller, *Chemical Exposures: Low Levels and High Stakes (2nd ed.)*, 36-37, 49-51, 83 (Wiley & Sons 1998).

[495] Nicholas Ashford & Claudia Miller, *Chemical Exposures: Low Levels and High Stakes (2nd ed.)*, 38-40 (Wiley & Sons 1998).

[496] Nicholas Ashford & Claudia Miller, *Chemical Exposures: Low Levels and High Stakes (2nd ed.)*, 36-37, 49-51, 83 (Wiley & Sons 1998).

drawal symptoms.[497] If they are away from the chemical too long—they will feel terrible when they are exposed again on Monday.

Rather than adapting to exposure through addiction, some of us will develop intolerance to chemicals in the same way some people become intolerant to nicotine, caffeine, and alcohol. We may by-pass the feel-good stage completely and move directly into the feel-bad stage after exposure. Chemical sensitivity is one form of chemical intolerance caused by chemical exposure (another form is chronic fatigue, discussed in the next chapter).[498] Indeed, people with a history of chemical exposure from chemical spills, pesticide applications, poor indoor air quality, and/or medical implants are prone to develop chemical sensitivity and chronic fatigue.[499]

Chemical sensitivity is a state in which the body becomes hyper-aware and highly intolerant to toxic chemicals. It is probably an adaptation strategy devised by the immune and nervous systems to encourage avoidance of toxic chemicals when the body cannot handle any more. Airborne chemicals—such as those off gassing from fragrance, plastics, paint fumes, car exhaust—are some of the worst for the chemically sensitive because a large amount of the toxic molecules quickly travel into the lungs and go directly into the bloodstream. As a result, the body quickly picks up on these substances and has an adverse reaction.

The liver of the chemically sensitive is likely compromised from chemical toxicity, undermining its ability to detoxify incoming substances. Any additional exposure will hit hard and will also result in strong withdrawal symptoms. It is probably for this reason that individuals who develop chemical sensitivity are also much more likely to develop drug intolerances, including intolerance to caffeine, alcohol, and even medications.[500] The body starts to react to any foreign substance that it perceives will burden the liver.

Chemically sensitive individuals also tend to be more sensitive to EMF than the general population. EMF is radiation from cell phones, cell towers, smart meters, Wi-Fi, etc. Both EMF and toxic chemicals cause damage

---

[497] Nicholas Ashford & Claudia Miller, *Chemical Exposures: Low Levels and High Stakes (2nd ed.)*, 42-44 (Wiley & Sons 1998).

[498] See e.g., Dunstan RH, et al., "Bioaccumulated chlorinated hydrocarbons and red/white blood cell parameters," *Biochem Mol Med*, 58(1), 77-84 (1996 Jun); and Nicholas Ashford & Claudia Miller, *Chemical Exposures: Low Levels and High Stakes (2nd ed)*, ch 1, 228–231, 233–260, 314 (Wiley & Sons 1998).

[499] See Claudia Miller & Thomas Prihoda, "A controlled comparison of symptoms and chemical intolerances reported by Gulf War veterans, implant recipients and persons with multiple chemical sensitivity." *Toxicology & Industrial Health*, 15(3–4), 386–397 (April 1999); see also Chapter 20 in this book, discussing chronic fatigue.

[500] Claudia Miller & Thomas Prihoda, "A controlled comparison of symptoms and chemical intolerances reported by Gulf War veterans, implant recipients and persons with multiple chemical sensitivity." *Toxicology & Industrial Health* 15(3–4), 386–397 (April 1999).

and stress to the cells and mitochondria.[501] The chemically sensitive person already has chemical damage to their cells and mitochondria (see Chapter 15). EMF threatens to cause additional stress and damage at the cellular level, and the chemically sensitive body is especially vulnerable to it. The chemically sensitive person will thus frequently experience EMF intolerance through symptoms such as headaches, inflammation, searing pain, etc.

Many chemically sensitive individuals do not identify with having a chronic illness. Nevertheless, they report that they feel fatigued after chemical exposure. A visit to a new building, industrial facility or perfumed space may make them feel very woozy and sleepy. Some report needing to take a nap later the same day. Many others feel exhausted and need bed rest for an entire day or two thereafter. Some also develop severe body aches and pains (fibromyalgia-like symptoms). Once they recover, life goes back to normal—until another exposure triggers a reaction.

> I had to sleep three solid days after a half hour exposure at work to damp carpets that had been cleaned with a chemical.

> —Anonymous Victim

A recent study found that the number of people who self-report as being chemically sensitive has doubled in the last decade to over 25% of the US population. The number of people who have actually been formally diagnosed as having chemical sensitivity has tripled in the last ten years to nearly 13%.[502] That is pretty substantial.

People's self-reported reactions to everyday chemicals include "headaches, dizziness, cognitive impairment, breathing difficulties, heart palpitations, nausea, mucous membrane irritation, and asthma attacks."[503] Such reactions are frequently triggered by "pesticides, new carpet and paint, reno-

---

[501] For more information on the health impacts of EMF pollution, see e.g., Alicja Bortkiewicz, et al., "Mobile phone use and risk for intracranial tumors and salivary gland tumors – A meta-analysis," *International Journal of Occupational Medicine & Environmental Health* 30(1), 27–43 (2017); Houston, BJ, et al., "The effects of radiofrequency electromagnetic radiation on sperm function," *Reproduction* 152(6), R263-R276 (Dec. 2016); Cindy Sage and David O. Carpenter (eds.) "Bioinitiative 2012: A Rationale for Biologically-based Exposure Standards for Low-Intensity Electromagnetic Radiation" BioInitiative Working Group (Dec. 2012); Carl Blackman, "Cell phone radiation: Evidence from ELF and RF studies supporting more inclusive risk identification and assessment," *Pathophysiology* 16(2-3), 205-216 (Aug 2009); Karl Hecht, et al., "Health Implications of Long-Term Exposure to Electrosmog," Competence Initiative for the Protection of Humanity, the Environment and Democracy, Brochure 6 (2017), available at https://www.emfanalysis.com/wp-content/uploads/2017/01/German-Report-on-878-Russian-EMF-Health-Studies.pdf.

[502] Anne Steinemann, "National Prevalence and Effects of Multiple Chemical Sensitivities," *Journal of Occupational and Environmental Medicine* 60(3): e152–e156 (March 2018).

[503] Anne Steinemann, "National Prevalence and Effects of Multiple Chemical Sensitivities," *Journal of Occupational and Environmental Medicine* 60(3): e152–e156 (March 2018).

vation materials, diesel exhaust, cleaning supplies, perfume, scented laundry products, and air fresheners."[504]

Some people will never develop a noticeable intolerance to chemicals, regardless of the dose. Our biological response will vary, depending in large part on how much exposure we have had, how frequently, to what types of chemicals, and in what order.[505] For each person, this will be different. And the details are likely to be unknown to most of us.

> Two different people will react differently to the same toxic substance because they have different immune systems, [different germs and vaccination histories and] different imbalances in their gut flora, for example. They may react differently because they already have had exposure to mercury from their dental amalgams, [exposure to] various antibiotics from pharma and farming operations, [have] organs with varying degrees of dysfunction due to injuries or nutritional deficiencies, etc.
>
> —Miranda Taylor, High Point Health [506]

Regardless of whether we become noticeably intolerant or not, long-term chronic exposure will ultimately lead to health problems—which can be large or small. And while we may not associate cause with effect, chronic health problems triggered by chemical exposure are a form of environmental illness.

---

[504] Anne Steinemann, "National Prevalence and Effects of Multiple Chemical Sensitivities," *Journal of Occupational and Environmental Medicine* 60(3): e152–e156 (March 2018).

[505] Claudia Miller, "Toxicant-Induced Loss of Tolerance – An Emerging Theory of Disease?," *Environmental Health Perspectives*, 105, supplement 2 (March 1997), republished and available online at http://www.tldp.com/issue/210/toxicantin.htm.

[506] Email from Miranda Taylor, Licensed Acupuncturist, Herbalist, & Nutrition Response Testing Practitioner at High Point Health in Seattle, dated February 25, 2020.

Chapter 22: Silent Winter

Some people develop the chemical flu after exposure. They usually believe that they have fallen ill with a virus or other infection because they first experience their deteriorating health as a set of flu-like symptoms.[507] These symptoms drag on, will not go away, keep coming back, or even get worse. They go through multiple rounds with doctors and medical practitioners of all kinds, being tested for Epstein-Barr, other viruses, Lyme disease, bacterial infection, mold, parasites, etc. Most are ultimately looked upon skeptically, blankly, or with exasperation by the conventional medical community while they try to cling on to their normal life.

One common health issue for those experiencing the chemical flu is chronic fatigue. Fatigue itself is arguably the most typical human response to any foreign invading agent (chemical, viral, bacterial, or other). When our body senses a dangerous substance at the cellular level, it starts to significantly shut down activity to protect itself from further exposure. The cells literally close down and stop letting substances in and out of them.[508] The body also gets busy dealing with the problematic substances in our system. When this happens, our energy levels become limited.

Chronic fatigue occurs when our bodies are overwhelmed by something for a much longer period of time. Fatigue is no longer a fleeting condition; rather, it becomes a state of being. The body enters a prolonged cellular

---

[507] Unger ER, Lin JS, Brimmer DJ, et al., "CDC Grand Rounds: Chronic Fatigue Syndrome— Advancing Research and Clinical Education," *MMWR Morbitity & Mortality Weekly Report* 65, 1434–1438 (Dec. 2016).

[508] See generally Robert K. Naviaux, "Metabolic Features of Cell Danger Response," *Mitochondrion* 16, 7-17 (May 2014).

defense mode, meaning that the cells are not letting anything move freely in and out of them, unfortunately also limiting access to nutrients and the elimination of toxins at a cellular level. The longer the state persists, the more severe the condition becomes. The body becomes increasingly starved of nutrients at the cellular level. The cells get increasingly overwhelmed by toxins already within their borders. This is an exhausting state for the body to be in for extended periods of time. In order to conserve energy under these lock-down conditions, the body enters a particular form of hibernation experienced as chronic fatigue.[509] A true state of Silent Winter.

Chronic fatigue caused by toxic chemicals begins with the same symptoms one would expect with an infection. They may include headache, fever, chills, nausea, cough, muscle aches, joint pain, rash, or fatigue.[510] This is because the body views toxic chemical exposure like any other infection or foreign invasion within the body.

> I was exposed to Carbamate pesticide when my house was sprayed for [bug] infestation soon after I had moved in...Thereafter, every time [we used] the vacuum cleaner [it] disturbed the remains of it. I reacted with flu like symptoms and extreme [allergic] reactions...
>
> —Anonymous Victim

> I was [diagnosed with chronic fatigue] in 1985. Abrupt onset [with] flu like symptoms. Periods of improved health and return to work. Never fully recovered but functional. Following treatment for cancer with chemotherapy and two years later a whole house renovation with chemical off gassing...I crashed and never recovered. That was eight years ago. I am now severely ill. House and bed bound. The decline of my cognitive function has been the worst part of it. I can no longer read or watch a movie. I was a documentary film producer.
>
> - Anonymous Victim

Flu-like symptoms resulting from chemical exposure are well-known to industry. For example, employees who worked with toxic PFAS chemicals at DuPont referred to having the "Teflon flu."[511] Fatigue, joint/muscle pain, cough and fever are known to be caused by "metal fume fever" in industrial

---

[509] See generally Robert K. Naviaux, "Metabolic Features of Cell Danger Response," *Mitochondrion* 16, 7-17 (May 2014).

[510] "Environmental Illness," Michigan Medicine, University of Michigan (Dec. 13, 2018), available at https://www.uofmhealth.org/health-library/zp3199; see also Chapters 19-20.

[511] Robert Bilott, *Exposure* (Simon & Schuster 2019).

workers.[512] Occupational exposure to formaldehyde is known to induce flu-like symptoms.[513] A flu-like epidemic has been induced in workers exposed to vinyl-based plastic fumes.[514] The list goes on.

> I went to work at Ford motor company at a production plant in 1976. 3 months later after chemical exposure, I came down with [chronic fatigue]. I know [that I] was exposed to lead. I know I was also exposed to PCB's....I had a doctor that thought I may [have] had Lupus. I tested negative. My sleep had [also] evaporated....Then in the late 80s I had the bad flu that I never recovered from. Eventually I developed Lymphoma and lived through that. I have lived with [chronic fatigue] for such a long time it is just my normal.
>
> —Anonymous Victim

A viral infection or other factor can be involved in the development of chronic fatigue or any other environmental illness. Initially chemical overload causes significant damage to our immune system, robbing us of our ability to effectively deal with potential viruses, bacteria, etc.[515] Subsequently an infection can become the final straw that breaks the camel's back. The invading bug can be relatively benign (like the cold virus) but become very hard to eliminate in the chemically injured. For this reason, long-term chronic fatigue sufferers sometimes refer to their disease as chemical AIDS.

> For 12 years I lived in a valley that did crop dusting....We woke up to the sound of the planes and most of the summer the air in the valley tasted like pesticides. We would even play outside while the planes passed overhead. The mist made the air sticky. But I didn't get sick [with chronic fatigue until] I came down with a severe case of Mono.
>
> —Anonymous Victim

> There are now four members of my family with [chronic fatigue]. We were exposed to a now-banned pesticide, chlordane...which had been used as a termiticide under our home before we bought it. [Three]

---

[512] "Occupational Diseases - A Guide to their Recognition," The National Institute of Safety and Health, 409 (1977), quoted in Letter from the Industrial Commission of Ohio to Owens-Corning Fiberglass Corp., dated April 1, 1986, available at toxicdocs.org.

[513] "Fact Sheet on Formaldehyde" Ford Motor Company (est. 1980), available at toxicdocs.org.

[514] ConocoVista memo re "Paragon Plastics Health Allegations," (May 13, 1991), available at toxicdocs.org.

[515] Jamie C. DeWitt, et al., "Exposure to per-fluoroalkyl and polyfluoroalkyl substances leads to immunotoxicity: epidemiological and toxicological evidence," *Journal of Exposure Science & Environmental Epidemiology* 29, 148–156 (Nov 2019).

of us became ill over the course of 3 years....Our fourth family member, our son, became ill several years later after an exposure to chemicals... [Three] of us appeared to have an infection (two with antibiotic use) immediately prior to suddenly becoming very ill, although we were being exposed for years. We have significant chemical sensitivity and still have [chronic fatigue] even though we are now living in a home built with safer materials.

—Anonymous Victim

Chronic fatigue is not just long-term fatigue. It involves a whole host of other symptoms signaling a dysfunction in the body. These can include weak or irregular heartbeat, dizziness, ear ringing, slow wound healing, brain-fog, loss of mental cognition, slurring of speech, nausea, temperature regulation issues, debilitating headaches, memory loss, neurological problems, muscular issues, and joint problems.[516]

Chronic fatigue is similar in many respects to a few other chronic diseases, including fibromyalgia and lupus. Fibromyalgia patients experience chronic fatigue. However, they also experience chronic pain in the body—in the muscles, ligaments, tendons and/or joints.[517] It is unclear whether chronic fatigue and fibromyalgia are two separate diseases or a slightly different manifestation of the same thing. For example, chronic fatigue used to be called myalgic encephalomyelitis (ME). "Myalgic" means "muscular pain" and "encephalomyelitis" means "inflammation of the brain and spinal cord."[518] The distinction between chronic fatigue and fibromyalgia is therefore quite blurry. Whether a patient is diagnosed with chronic fatigue or fibromyalgia may depend on the diagnosing doctor and the symptoms the patient may be experiencing at the time.

Lupus, a rarer disease that affects African Americans and women disproportionately, is also quite similar to chronic fatigue and fibromyalgia. Lupus is essentially chronic fatigue and significant body pain with a characteristic butterfly rash on the face.[519] While lupus is most commonly associated with a specific type of inflammation, research confirms that inflammation

---

[516] Based on interviews with chronic fatigue sufferers; see also Mangalathu S.Rajeevan, "Pathway-focused genetic evaluation of immune and inflammation related genes with chronic fatigue syndrome," *Human Immunology* 76(8), 553-560 (Aug 2015).

[517] "Symptoms," National Fibromyalgia Association, available at http://www.fmaware.org/wp-content/uploads/2017/05/Symptoms.pdf (last visited May 22, 2019).

[518] Merriam Webster Dictionary Online, available at https://www.merriam-webster.com/dictionary/encephalomyelitis and https://www.merriam-webster.com/dictionary/myalgic (last visited May 22, 2019).

[519] "Lupus - Overview," Mayo Clinic, available at https://www.mayoclinic.org/diseases-conditions/lupus/symptoms-causes/syc-20365789 (last visited May 22, 2019).

also goes hand in hand with fibromyalgia[520] and chronic fatigue in general.[521] These diseases seem to ultimately stem from the same types of problems but are expressed differently in different people.

Recent estimates suggest that up to 4 million people in the US suffer from chronic fatigue syndrome.[522] Another 10 million people are estimated to have fibromyalgia (chronic fatigue + joint pain).[523] Lupus (chronic fatigue + joint pain + rash) is believed to affect under one million Americans. Collectively chronic fatigue, fibromyalgia and lupus affect nearly 5% of the US population or 1 in 20 people. These numbers are estimates only, as researchers believe that up to 85% of US citizens suffering from chronic fatigue may never be diagnosed.[524] This is because the disease is poorly acknowledged in society and because many people fear that a formal diagnosis would adversely impact their future career opportunities and health insurance benefits.

Gulf War veterans have some of the highest incidences of chronic fatigue—20% have been diagnosed with either chronic fatigue or fibromyalgia (i.e. fatigue + joint pain).[525] They further are known to have digestive problems and chronic skin conditions, among others.[526] The prevalence of chronic fatigue in Gulf War veterans is so high that Gulf War Veterans who develop chronic fatigue are automatically eligible to receive VA disability compensation."[527] The US government concedes that Gulf War veterans

[520] Coskun Benlidayi, "Role of inflammation in the pathogenesis and treatment of fibromyalgia." *Rheumatology International* 39(5), 781-791 (Feb 2019).

[521] Mangalathu S.Rajeevan, "Pathway-focused genetic evaluation of immune and inflammation related genes with chronic fatigue syndrome," *Human Immunology* 76(8), 553-560 (Aug 2015).

[522] Ashley R. Valdez, "Estimating Prevalence, Demographics, and Costs of ME/CFS Using Large Scale Medical Claims Data and Machine Learning," *Frontiers in Pediatrics* 6, 412 (Feb 2018); Peggy Rosati Allen, "Chronic Fatigue Syndrome: Implications for Women and their Health Care Providers During Childbearing Years," *Journal of Midwifery and Women's Health* 53(4), 289-301 (Jul-Aug 2008).

[523] "Symptoms" & "Prevalence," National Fibromyalgia Association, available at http://www.fmaware.org/about-fibromyalgia/symptoms/ and http://www.fmaware.org/about-fibromyalgia/prevalence/ (last visited May 22, 2019).

[524] Peggy Rosati Allen, "Chronic Fatigue Syndrome: Implications for Women and their Health Care Providers During Childbearing Years," *Journal of Midwifery and Women's Health* 53(4), 289-301 (Jul-Aug 2008) and Testimony of Dr. Suzanne Vernon, Centers for Disease Control Press Briefing Transcripts, "Chronic Fatigue Syndrome," dated April 20, 2006, available at https://www.cdc.gov/media/transcripts/t060420.htm.

[525] J.W. Ashford, "Caring for ODS/S Veterans at the WRIISC: Focus on Symptoms: Chronic Pain, Chronic fatigue, Irritable Bowel Syndrome, etc.," US Department of Veterans Affairs (June 25, 2018), available at https://www.va.gov/RAC-GWVI/meetings/jun2018/AshfordRAC20180625508compl.pdf.

[526] Eisen SA, et al., "Gulf War veterans' health: medical evaluation of a U.S. cohort," *Annals of Internal Medicine* 142(11), 881-890 (Jun 2005).

[527] "Myalgic Encephalomyelitis/Chronic Fatigue Syndrome in Gulf War Veterans," US Department of Veteran's Affairs, available at https://www.publichealth.va.gov/exposures/gulfwar/chronic-fatigue-syndrome.asp (last visited April 15, 2019).

have had a variety of chemical exposures to pesticides, chemical warfare, radioactive materials, contaminated water, oil well smoke, jet fuels, solvents, hydraulic fluids and other toxic petrochemicals, that individually or collectively has caused their physical illnesses and disabilities.[528]

Military recruits at home are facing similar issues. The majority of our military bases within the US are contaminated with toxic chemicals from various training operations.[529] For example, one of the most contaminated army bases is located at Fort McClellan in Alabama. PCB's, nerve gases, radioactive compounds, Agent Orange, and other chemicals are found there, poisoning the locals.[530]

> I had basic training at Ft McClellan Alabama....many there are fighting to get the government to recognize that our health issues stem for exposure to toxic chemicals while there....
>
> —Anonymous Chronic Fatigue Victim

PFAS, a toxic class of chemicals frequently used as flame retardants (see Chapters 11–12), have recently been added to the list as a result of widespread contamination at US military facilities.[531]

Citizens who have served our country in other ways have also developed chronic fatigue as a result of toxic chemical exposure. For example, volunteers who stepped up during 9/11 at Ground Zero of the World Trade Center suffer a variety of chemical-related illnesses, including chronic fatigue.

> [My sister and I] were both volunteers down at the World Trade Center at Ground Zero. We were citizens who responded to the attack on the World Trade Center. I volunteered through the Red Cross and then the Salvation Army....Subsequent to my stepping foot at Ground Zero I have developed, in this order, vitiligo, which is a disease of the skin pigmentation; asthma and reactive airway disease; post-traumatic stress, depression and generalized anxiety disorder; GERD and

---

[528] J.W. Ashford, "Caring for ODS/S Veterans at the WRIISC: Focus on Symptoms: Chronic Pain, Chronic fatigue, Irritable Bowel Syndrome, etc.," US Department of Veterans Affairs (June 25, 2018), available at https://www.va.gov/RAC-GWVI/meetings/jun2018/AshfordRAC20180625508compl.pdf.

[529] John W. Hamilton, "Contamination at U.S. Military Bases: Profiles and Responses," *Stanford Environmental Law Journal* 35(2), 223-249 (June 2016).

[530] "Potential Exposure at Fort McClellan," US Department of Veterans Affairs, available at https://www.publichealth.va.gov/exposures/fort-mcclellan/ (last visited July 10, 2019); and "Fort McClellan Groundwater Contamination Veterans Benefits," Woods & Woods LLC, available at https://www.woodslawyers.com/fort-mcclellan-groundwater-contamination-veterans-benefits/ (last visited July 10, 2019).

[531] Tara Copp, "DoD: At least 126 bases report water contaminants linked to cancer, birth defects," *Military Times* (April 26, 2018) and Melanie Benesh, "Mapping PFAS Chemical Contamination at 106 U.S. Military Sites," *Environmental Working Group* (March 6, 2019).

IBS; moderate obstructive sleep apnea; fibromyalgia; autoimmune antiphospholipid antibody syndrome, which is a blood disorder; lupus and Hashimoto's Disease, along with chronic fatigue....[532]

The reason why Ground Zero was so toxic is because we construct buildings with materials containing highly problematic chemicals. These include the chemicals found in glues, coatings, paints and other synthetic or treated materials. All of these burned, combusted and volatilized into the air during the 9/11 attack. First responders rushed in and breathed in the fumes and fine particles from the ruins.

Office workers have also become sick from our built environment. The problem has become sufficiently common that it has a name: "Sick Building Syndrome." In 1994, the newly constructed DuPage County Courthouse outside of Chicago made national headlines when 700 people had to be evacuated as a result of the chemical off gassing.[533] A few of the employees at the courthouse were interviewed on the CBS evening news about their experience with chemically induced chronic fatigue:

> Mrs. Lori Chassee (Employee): "So we talked about the 3:00 nap time, people asleep at their desks, inability to wake up. But [we'd] do it almost tongue and cheek and [there was] a lot of laughing about it and laughing about the people that [were the] sickest."

> Ms. Claudette Lewis (Employee): "On the weekends, I was fine. I could function. And then on Monday morning, you'd go back and [the fatigue would] start all over again."

> Reporter: "Claudette Lewis has come down with chronic fatigue syndrome...."[534]

The 1980s and 1990s were particularly bad for the building industry. Many toxic chemicals and mold-prone building materials were used in construction. Operable windows went out of fashion and a "tighter building envelope" with little natural ventilation was all the rage. As a result, toxic fumes off-gassed into the building, rather than outside. A number of studies—including one study performed by the EPA itself—demonstrated that the

---

[532] Testimony of Denise Villamia to the US Department of Health and Human Services found in the transcripts of the "James Zadroga 9/11 Health and Compensation Act of 2010 Public Meeting" convened by the Department of Health and Human Services on March 3, 2011, at page 155:14-23.

[533] Jan Ferris, "Jury Blames DuPage for 'Sick' Courthouse," *Chicago Tribune* (Dec 31, 1994).

[534] Transcripts of CBS Evening News from October 12, 1992, as obtained from the Chemical Fabrics and Film Association, available at https://toxicdocs.org. The original video clip is available at https://tvnews.vanderbilt.edu/programs/341246.

indoor air quality was significantly worse than the outdoor air quality at such buildings.[535]

Indoor air quality can be a serious issue any time you introduce toxic chemicals into the indoor environment. A man I interviewed spent his childhood continuously exposed to toxic fumes from dry-cleaning chemicals and mothballs in the family's coat closet. This led to allergies and brain fog early on in life. The exposures also lay the foundation for health problems later on. He developed chronic fatigue after moving into a brand new building for work in his early adult life. He remembers reacting to the carpets and other new building materials. He first became sick with recurrent infections before falling down with chronic fatigue.

Research over the last decades confirms that people who fall ill with chronic fatigue generally have had multiple and cumulative exposures—all of which came at the wrong time for them.[536] Some are continuously around toxic substances at work. Others are inadvertently exposed to pesticides near their home. Those who are lucky can account for their major exposures. Many others may have some suspicions but do not truly know what was in their environment and when.

> [I worked at a nuclear energy plant when I was young and] by noon hour one day, my nylons had dissolved off of my legs. Now what was in the air that day, and did it bother anybody? I don't know....I got chronic fatigue syndrome in 1991. My younger sister of a year [who also worked at the nuclear energy plant] got it five years ahead of me.... I've got two aunts, one was 108 and one was 106, and I'm still a bit of a pistol. [But] I was in bed for three years -- '91, '92 and '93....[537]
>
> —Anonymous Victim

> My [chronic fatigue] became severe when I moved into a new timber-framed house which was heavily treated with [insecticides and wood preservatives]. I and a number of my [neighbors]...became ill with various immune illnesses. My young next door [neighbor] died, one other young woman, like me, ended up in a nursing home....I learnt only recently that a cousin's husband was the manager of the company

---

[535] Nicholas Ashford & Claudia Miller, *Chemical Exposures: Low Levels and High Stakes (2nd ed.)*, 11-15 (Wiley & Sons 1998).

[536] Robert Naviaux, "Metabolic Features of Cell Danger Response," Mitochondrion 16, 7-17 (May 2014).

[537] Transcript of testimony of Former Employee at Atomic US Facility given at the "Energy Employees Occupational Illness Compensation Program Act - Special Exposure Cohort Townhall Meeting #3," held by the National Institute for Occupational Safety & Health on August 7, 2002, at page 80:5-12.

who built the houses and so was regularly in contact with the treated wood. It caused him to become seriously ill and he died....

—Anonymous Victim

I grew up with a strawberry farm that bordered my backyard fence in the 1970s and early 80s. I've just recently discovered that the pesticides applied to this crop are particularly toxic.

—Anonymous Chronic Fatigue Sufferer

Those with chronic fatigue also develop intolerance to things such as household chemicals, fragrances, certain foods, sounds, mold, bug bites, Wi-Fi / EMF, and a whole host of other things. Usually the longer their state of chronic fatigue, the more severe the issues become. Chronic fatigue sufferers are especially chemically sensitive on average, compared to the rest of the general population.[538] Chemical sensitivity is likely the first indication of toxic chemical overload in their bodies that can ultimately result in long-term chronic fatigue or another chronic illness. Indeed, some level of chemical sensitivity is likely to develop years before other problems set in. As a result, chronic illness and multiple chemical sensitivity go hand in hand.

I've been diagnosed with Chronic Fatigue Syndrome, and yes, toxic exposures trigger me.

—Anonymous Victim

I've been diagnosed with fibromyalgia for a year and didn't know it was all connected. Just walking in a supermarket down the [laundry detergent aisle] is a nightmare [and] I try my best to hold my breath.

—Anonymous Victim

Industry has applied their lobbying efforts and PR magic to keep environmental illnesses such as chronic fatigue and chemical sensitivity from getting serious recognition.[539] If these diseases—along with other chronic illnesses—were publicly recognized as being caused by toxic chemicals rather than lifestyle, then they would become a liability for industry. We would recognize that toxic chemicals—rather than the victim—has caused

---

[538] Iris Bell, et al., "Illness from Low Levels of Environmental Chemicals: Relevance to Chronic Fatigue Syndrome and Fibromyalgia," *The American Journal of Medicine* 105(3A) Supp, 74S-82S (Sept. 1998).

[539] Anne McCampbell, "Multiple Chemical Sensitivities Under Siege," *Townsend Letter for Doctors and Patients* 210 (January 2001) (reprinted).

the problem and industry might be sued or even regulated. This does not sit well with industry (or our government, who does not like to regulate). As a result, the connection between toxic chemicals and environmental illness is frequently rejected, denied, ignored, and sometimes even treated as a form of emotional psychosis in public discourse. There are also public efforts to explain chronic illness away on a biological level excluding the link to toxic chemical exposure. Many studies seek to identify a common medical framework for talking about diseases without having to mention the role of chemical toxicity. Indeed, the pharmaceutical industry seeks to develop biomarkers for disease (rather than examining their environmental causes) so that they can develop responsive drugs and monetize a cure.

Industry has also worked very hard to silence the voices of sufferers of environmental illness so that they do not have to deal with them. One way has simply been to ignore the diseases, so that people who fall ill will simply go away. This strategy is so successful in occupational health that it falls under the rubric of: "the healthy worker effect." The relationship between chronic fatigue and the healthy worker effect was described back in the early 1980s in a Union Carbide internal corporate memo:

> If a worker...develops symptoms such as chronic fatigue or vague aches and pains, he may voluntarily "drop out." He often leaves a demanding job in big industry for a less demanding job in a small service industry. He will probably never get studied [in a worker health study]. Some 'drop outs' never get steady employment again.[540]

When industry has not been able to get victims of environmental illness to quietly go away, it has resorted to social ostracism such as harassment, lay-offs, stigmatization, on-the-job monitoring and the withholding of occupational diagnosis and treatment.[541]

[540.] Communication from T.A. Lincoln to Bonnie Almond, The "Healthy Worker Effect," Nuclear Division News (3/23/81), an internal Union Carbide document available at toxicdocs.org.

[541] Tamara L. Mix, "Social Control and Contested Environmental Illness: The Repression of III Nuclear Weapons Workers." *Human Ecology Review* 16(2), 172-183 (Winter 2009).

PART III: GETTING OUT OF THIS MESS

Chapter 23: Regaining Our Senses

> The mistake you make, don't you see, is in thinking one can live in a corrupt society without being corrupt oneself.
>
> —George Orwell

> It has long been suspected that many [collapses in civilization] were at least partly triggered by ecological problems; people inadvertently destroying the environmental resources on which their societies depended.
>
> —Jared Diamond[542]

Researchers have studied how we behave when facing a crisis. It turns out that all of us go through certain inevitable steps. At first we are in denial and may even fall into a rage at the suggestion that we need to pay attention. Eventually things get noticeably bad enough that we progress into a state of panic. At this phase, we tend to run around putting our energies into the wrong things, operating under an outdated paradigm of reality. This goes on until we either run out of energy and perish or are able to accept our change in circumstance. Those who take the latter route are able to problem-solve and find their way out.[543] Acceptance of our circumstances is essential for

---

[542] Jared Diamond, *Collapse: How Societies Choose to Fail or Succeed*, 6 (Viking Penguin 2005).
[543] Laurence Gonzales, *Deep Survival* (W.W. Norton & Company 1998).

finding our way out of our environmental crisis as well. It means letting go of the outdated beliefs and assumptions that comprise our current mindset.[544]

The root cause of our environmental problems is a mindset that puts profit over human health and environment. It permeates our decision-making patterns—both as individuals and collectively through the institutions that run our economy. Our current mindset it is built upon a paradigm of resource depletion through continuous colonization of our planet. In our colonizing world we can only exploit or be exploited. Each of us prefers to stay ahead, and so we act accordingly. We make decisions out of fear of being exploited and out ambition to exploit. We fear losing money, our jobs, our professional reputation, etc. We aspire to succeed within this system by gaining status, wealth, and social acceptance. We are constantly chasing or protecting our piece of the pie. We are so focused on these goals that we are willing to ignore the consequences and the inevitable victims of this cut-throat approach to the world.

Indeed, a significant proportion of the population learned that they could achieve a substantial measure of personal and economic success by shutting down emotionally and refusing to see our problems. This out-of-touch demographic usually views our social, environmental, and economic challenges as being caused by the personal weaknesses of others—that is, the victims are perceived as lazy or stupid. They deserved their fate or they must be emotionally unstable to claim that a problem exists.

The out-of-touch demographic closely follows the rules set forth by our paradigm of economic colonization. Because they have done pretty well for themselves working within the system, it is difficult for them to comprehend that one can follow the rules and still be suffering. A number of intellectuals have described this demographic as the Dreamers—those who believe the American Dream is real and that they are living it. The Dreamers do not intellectually acknowledge that they are living their socio-economic reality at the expense of others.[545]

Another significant portion of the population recognizes the problems within our society. However, they are so focused on maintaining harmony with others that they tend to emotionally deny the existence of the problem in their social and professional interactions. This demographic is fixated on

---

[544] Laurence Gonzales, *Deep Survival* (W.W. Norton & Company 1998); Carol S. Dweck, *Mindset*, (Ballantine Books 2006); Joanna Malaczynski, "Entrepreneurship Tip #1: Update Your Mental Model," DESi Potential (Aug 20, 2018); Joanna Malaczynski, "What I Learned About the Innovation Process from Practicing Law," *Innovation Excellence* (January 20, 2019).

[545] See e.g., Ta-Nehisi Coates, *Between the World and Me* (Spiegel & Grau 2015); and Toni Morrison, *The Origin of Others* (Harvard University Press 2017).

making their peers and the powerful people happy in order to feel socially and financially secure. They avoid pointing out uncomfortable facts, delivering bad news, or speaking out with an independent voice. Just like in the Hans Christian Andersen tale, *The Emperor's New Clothes*,[546] this latter group of individuals is very eager to please. They say and object to very little because they are afraid to be seen as a threat or as incompetent—by their boss or even the lady at the grocery store. They may object in their private lives and on social media, but they do not have the courage to object to our chemical world in the public sphere. They are too afraid of the consequences of doing so.

Each of us has utilized one or both of these emotional strategies throughout the course of our lives. They work well in a colonizing world, which encourages us to disassociate from ecological destruction and climb within the hierarchy. We adopt these strategies because we are dependent upon our economic system of colonization. The institutions who uphold this unfortunate regime (industry and government) are the source of our food, housing, health and tangible necessities. We are willing to do what it takes to keep them sufficiently happy so that they will provide us with income to buy the things that we need and desire. As a result, what we think, say, and do is generally out of alignment with how we feel.[547] This misalignment between our emotions and our reality is the source of our personal corruption and collective dysfunction as a species.

Our existing cultural mindset is based on a very specific emotional paradigm cultivated within modern society: Caring emotions—those that motivate us to preserve life and each other—have been marginalized into our private lives. Fear-based emotions—those that motivate us to take more than our fair share out of fear of scarcity—are encouraged in our public lives. Indeed, we are told that being fear-based (i.e., taking more than our fair share) is "logical" and appropriate in business. And that being caring—which does not maximize productivity but rather helps preserve life—is "irrational" and has no place in our economy.

Having drunk this cultural Kool-Aid, we suppress our caring emotions on an individual and global scale. The end result is that we approach the entire planet—its resources, life, and people—as something to be colonized and monetized. The goal is to make a profit by capitalizing on anything and

---

[546] Hans Christian Andersen, "The Emperor's New Clothes," *Fairy Tales Told for Children - First Collection* (C.A. Reitzel 1837).

[547] See e.g., Sarah Gibbons, "Empathy Mapping: The First Step in Design Thinking," Nielsen Norman Group (January 14, 2018), available at https://www.nngroup.com/articles/empathy-mapping/; Teal Swan, "The Mirror Event," Basel, Switzerland (2018), available at https://tealswan.com/premium/.

everything we can. We may not do this consciously, but we have codified this mantra by our individual refusal to feel the adverse impacts of our behavior.

Because we have publicly become shut off from our emotional reality, ecological problems are not seen, recognized, or even talked about when it matters. And solutions never have a chance to develop, let alone be discussed. If our employer is depleting natural resources and poisoning our communities, most of us do not complain because we are afraid of losing our job. If the powerful are engaging in activities that are destroying our ecosystems and undermining our health, most of us stay silent because our livelihood depends on their blessing. As a result, the status quo gets stronger, while our health and environment get weaker and sicker.

On a day-to-day basis, we practice alienating from our emotions in order to aspire to the American Dream of economic abundance. If we just toe the line, we will make it. If we buy low and sell high, we will come out ahead. What we do not recognize, however, is that we are not living the American Dream. Instead, we are destroying our health and environment, believing that we are making progress for ourselves.

By suppressing conflicting emotions around our health and environmental issues, we become complicit in the environmental problems we see around us. We have culturally learned to ignore these problems; it is the equivalent of sticking our head in the sand. It is not possible to problem-solve from such a position of denial. And we need to be brave enough to reawaken our emotional senses in order to get out of this predicament.

Chapter 24: Knowing What Is Possible

> We live our lives, grateful that things aren't worse than they are. But there has to be a threshold beyond which we can no longer ignore the destructiveness of our way of living.... [And it is] up to us to determine for ourselves how closely the patterns we've been handed by our culture fit our experience of the world.
>
> —Derrick Jensen[548]

Industry almost always claims that sustainable technologies do not exist, in order to justify their toxic products and resulting environmental pollution. Alternatively it claims that alternatives are not good enough or not scalable in the marketplace. We buy these arguments because we do not see sustainable technologies in the marketplace. We do not realize, however, that there is little or no incentive for companies to adopt sustainable technologies in the first place.[549]

Based on my professional experience, I can attest that sustainable alternatives exist everywhere. However, sustainable technologies have difficulty finding corporate support in the marketplace.[550] This is because industry is

---

[548] Derrick Jensen, *A Language Older Than Words*, 15, 42, (Context Books 2000).

[549] See e.g., Joanna Malaczynski, "Why Your Angel Investment is Going to the Status Quo," DESi Potential (November 19, 2019); Joanna Malaczynski, "Advice for Innovative Companies Entering Emerging Markets," *Young Upstarts* (July 26, 2018); Joanna Malaczynski, "Three Customer Discovery Tips for Building an Innovative Company," LinkedIn Articles (July 16, 2018); Joanna Malaczynski, "Three Lessons for Building a Successful Company with Sustainability Benefits," *Young Upstarts* (July 10, 2018).

[550] See e.g., Jay Harman, *The Shark's Paintbrush* (White Cloud Press 2013); Paul Hawken, et al., *Natural Capitalism* (Little, Brown & Company 1999); and Chapter 1.

not interested in sustainable ways of doing things in the absence of regulation, despite what their marketing materials might claim. They generally perceive that implementing alternatives would take too much effort. Indeed, even the more progressive companies out there have only taken half-hearted steps to invest in safer alternatives to toxic chemicals.

Absent regulatory pressure, there is little motivation to change. And there is a lot of push-back from the status quo companies that are happy making billions on their existing product lines. It is easy for the establishment to convince others (and themselves) that safer substitutes do not exist. And companies are not trying very hard to find them. I have repeatedly witnessed corporate employees falsely or erroneously claim that there are no safer substitutes for their technologies. In one instance, the manufacturer of the safer substitute was in the very same room.

Industry needs to shift its energies away from resisting change to the implementation of greener and healthier solutions. To make this happen, we need to set hard and fast deadlines for them to change. We further we need to provide industry with support to meet regulatory deadlines by investing corporate tax dollars into the implementation of sustainable technologies. Regulatory agencies can be utilized very effectively to help industry convert from being environmental polluters to sustainable companies. Many regulatory agencies already have experience working with industry to find safer alternatives to toxic chemicals.[551] These efforts need to be ramped up exponentially, backed by corporate tax dollars, and made absolutely mandatory. Otherwise it will all be a pipe dream and industry will fight us tooth-and-nail along the way.

For companies whose very products are inherently toxic—such as pesticide manufacturers—the focus needs to be on transitioning into sustainable economic activity based on their technical and economic expertise. This means delegating to them the task of directly cleaning up the mess they have made and developing truly sustainable economic practices. Counter-intuitively, we need to support these companies in their transition to sustainability. Their task should not be a financial punishment but rather an economic journey toward success. We need these companies (and all of their employees) to come out the other end as providers of legitimate economic

---

[551] See e.g., "Pollution Prevention Assistance ," State of Washington Department of Ecology (last visited February 21, 2020); "Mission, Vision & Values," Northwest Green Chemistry (last visited February 21, 2020); "Pollution Prevention" and Minnesota Technical Assistance Program," Minnesota Pollution Control Agency (last visited February 21, 2020); "About," Toxics Use Reduction Institute (TURI) (last visited February 21, 2020); "Mission, Vision & Values" and "About IC2" Interstate Chemicals Clearinghouse (IC2) (last visited February 21, 2020).

services. Otherwise they will wage war against society in order to survive and maintain their status quo.

Ultimately corporations need to shift their resources to investing in ecological abundance in order to create a healthy economy. A sustainable civilization—be it ancient or modern—recognizes that we are ultimately dependent upon nature for our survival. Nature is the source of wealth; not the financial system, not industry, not government, and not colonization. We historically understood the importance of ecological abundance because we directly relied on nature for survival. No fish, no food. No clean water, no fish. We also recognized that nature does not consume and exhaust our essential resources; rather, nature generates them.

Our records still document the abundance of nature, as formerly found in ecosystems across the globe. For example, Europeans who first came to North America were astounded that the rivers were filled to the brim with fish and vast fields were covered with more berries than anyone could ever harvest.[552] This ecological abundance is what allowed—and still allows—civilization to thrive. We are out of touch with this reality because we no longer see the direct connection between our health, our wealth and nature.[553] It is still there, however, as unavoidable as death itself.

One way for industry to invest in nature is by applying the principles of biomimicry to both manufacturing and business operations. Biomimicry is the science of how to emulate nature to create more sustainable and resilient products, technologies, companies and economies.[554] Nature has taken billions of years to come up with the most advanced technologies known on the planet. We have a lot to learn from them. Nature builds systems that are self-sustaining, self-perpetuating, and resilient. Nature has also created chemistries, sources of energy, and complex systems using benign technologies that are generally life-giving. Biomimicry has already been used to create many successful technologies: from safer high-performance adhesives to non-toxic anti-microbial surfaces.[555] It is a way to design and build a sustainable world, applying life-giving methodologies.

---

[552] William Cronon, *Changes in the Land*, Hill & Wang (1983).

[553] Jared Diamond, *Collapse: How Societies Choose to Fail or Succeed* (Viking Penguin 2005).

[554] See e.g., Janine Benyus, *Biomimicry: Innovation Inspired by Nature* (Harper Collins 2002); Jay Harman, *The Shark's Paintbrush* (Nicholas Brealey Publishing 2013). For more information, see e.g., "Biomimicry Basics Online Course," and the AskNature database from the Biomimicry Institute.

[555] See e.g., Janine Benyus, *Biomimicry: Innovation Inspired by Nature* (Harper Collins 2002); Jay Harman, *The Shark's Paintbrush* (Nicholas Brealey Publishing 2013). For more information, see e.g., "Biomimicry Basics Online Course," and the AskNature database from the Biomimicry Institute.

One secure way for us to lay the foundation for such sustainable design and investment is to pass a constitutional amendment that formalizes our commitment to "a healthy environment that does not cause chronic disease."[556] But other interventions are needed. We do not have effective environmental laws because industry lobbies against them. Industry is able to do that successfully when it gains too much market power. Our antitrust laws were intended to prevent companies from amassing market power. However, these laws have consistently been eroded by industry since their inception. These laws also never anticipated that companies would conspire to gain economic power at the expense of our health by suppressing scientific information. Nor that industry would collude with our government and amass influence internationally. We need to significantly update our antitrust laws to address these exploitative behaviors.

We also need stronger transparency requirements, forcing companies to disclose to us what substances are being used and their impacts on our health. It should not be the government's burden to parse out what is safe versus unsafe under economic pressure to provide a stamp of approval. Nor should the onus be on small businesses to certify to us that their products are non-toxic, organic, GMO-free, phthalate-free, etc. Rather, the onus should be on the status quo to take responsibility for the safety of their products by making chemical health hazards transparently clear to everyone around.

---

[556] Naviaux Lab, "The 28th Amendment Project," available at naviauxlab.ucsd.edu/the-28th-amendment-project/ (last visited February 19, 2020).

ACKNOWLEDGEMENTS

This work was based on the efforts of many authors who have come before me, including Rachel Carson, Theo Colborn, Claudia Miller, Carol Van Strum, E. G. Vallianatos, Shiv Chopra, Carey Gillam, Rob Bilott, Vandana Shiva, and those who came before them. It was also made possible by the countless citizens who have fought for clean air, soil, food, and water in their communities and by the witnesses who have documented their struggle. It was also made possible by the individuals who communicated with me about their environmental illness.

My capacity to write this book was shaped by my mentors and peers. I am especially grateful to the lawyers, scientists, designers, and others who have over the years helped me understand the relationship between toxic chemicals and human health. This includes Mark Todzo, Eric Somers, and Howard Hirsch at Lexington Law Group, the members of the Interstate Chemicals Clearinghouse, as well as corporate and regulatory staff I befriended who I hesitate to mention in person due to the critical nature of this book toward industry; you know who you are. I also extend my deepest gratitude to every one of my mentors when I was a practicing antitrust attorney: Paul Alexander, Stephen Bomse, and Larry Popofsky—your mentorship was particularly foundational in my ability to articulate and analyze the ways in which companies affect and influence the marketplace.

I received much support from my family along my journey. Thank you for your encouragement and for making it possible for me to have the space and time to write. I especially wish to thank my husband, Timothy Moore, for believing in me, listening to my thoughts about the book, and being a loving force in my life.

Thank you to Algora Publishing for the essential work of getting my writing into your hands, including Martin DeMers for making this book a priority. Thank you also to my editor, Andrea Secara, for her thoughtful support throughout the publishing process.

Finally, thanks to those of you who will read this book and through your choices shift the momentum of our world into a better future.

Printed in the United States
by Baker & Taylor Publisher Services